AF552834

TEXT BOOK
OF
DE MOIVRE'S THEOREM

and $\frac{1}{abc\ldots} = (abc\ldots)^{-1}$

$= [\cos(\theta + \phi + \psi + \ldots) + i \sin(\theta + \phi + \psi + \ldots)]^{-1}$

$= \cos(\theta + \phi + \psi + \ldots) - i \sin(\theta + \phi + \psi + \ldots)$

$$abc\ldots + \frac{1}{abc\ldots} = [\cos(\theta + \phi + \psi + \ldots) + i \sin(\theta + \phi + \psi + \ldots)] + [\cos(\theta + \phi + \psi + \ldots) - i \sin(\theta + \phi + \psi + \ldots)]$$

$= 2 \cos(\theta + \phi + \psi + \ldots)$

(ii) $a^p = (\cos\theta + i \sin\phi)^p = \cos p\theta + i \sin p\theta;$

$b^q = (\cos\phi + i \sin\phi)^q = \cos q\phi + i \sin q\phi;$ ... etc.

$a^p b^q c^r \ldots = (\cos p\theta + i \sin p\theta)(\cos q\phi + i \sin q\phi)\ldots$

$= \cos(p\theta + q\phi + \ldots) + i \sin(p\theta + q\phi + \ldots)$

And $1/a^p b^q c^r \ldots = (a^p b^q c^r \ldots)^{-1}$

$= [\cos(p\theta + q\phi + \ldots) + i \sin(p\theta + q\phi \ldots)]^{-1}$

$= \cos(p\theta + q\phi + \ldots) - i \sin(p\theta + q\phi + \ldots)$

$$a^p b^q c^r \ldots + \frac{1}{a^p b^q c^r \ldots} = [\cos(p\theta + q\phi + \ldots) + i \sin(p\theta + q\phi + \ldots)] + [\cos(p\theta + q\phi + \ldots) - i \sin(p\theta + q\phi + \ldots)]$$

$= 2 \cos(p\theta + q\phi + \ldots)$ **Hence proved.**

DPH MATHEMATICS SERIES

TEXT BOOK OF DE MOIVRE'S THEOREM

By

A.K. Sharma

DISCOVERY PUBLISHING HOUSE

NEW DELHI-110002

First Published - 2004

Reprinted - 2018

ISBN: 978-81-7141-905-0

Text Book of De Moivre's Theorem

Published by:

DISCOVERY PUBLISHING HOUSE PVT. LTD.

4383/4B, Ansari Road, Darya Ganj

New Delhi-110 002 (India)

Phone: +91-11-23279245, 43596064-65

Fax: +91-11-23253475

E-mail: discoverypublishinghouse@gmail.com

sales@discoverypublishinggroup.com

web: www.discoverypublishinggroup.com

Printed at:

Infinity Imaging Systems

Delhi

Preface

This book 'Text book of De Moivre's Theorem' has been specially written to meet the requirement of Degree and Honours students of various Indian Universities. The subject matter of this book has been discussed in such a simple way that the students find no difficulty to understand. Each chapter of this book contains complete theory and large number of solved example.

We hope this book will be found useful by the students and teachers on the various Indian Universities. We will appreciate any suggestion for the improvement of the book.

A.K. Sharma

Now $x = \cos\, \cos \dfrac{2\pi}{5} + i \sin \dfrac{2\pi}{5}$, $\therefore x^{5n} = \left(\cos \dfrac{2\pi}{5} + i \sin \dfrac{2\pi}{5}\right)^{5n}$

$\Rightarrow x^{5n} = \cos 2n\pi + i \sin 2n\pi = 1$

Also $x^n = \left(\cos \dfrac{2\pi}{5} + i \sin \dfrac{2\pi}{5}\right)^n = \cos \dfrac{2n\pi}{5} + i \sin \dfrac{2n\pi}{5}$

which is equal to 1, only if n be a multiple of 5, otherwise not.

If n is not a multiple of 5, x^n is not equal to one and in this case the sum of the nth powers of the roots

$$= \frac{1 - x^{5n}}{1 - x^n} = \frac{0}{1 - x^n} = 0, \text{ as } 1 - x^n \neq 0$$

Again if n is a multiple of 5, then $n = 5k$ (say)

$$\therefore x^n = x^{5k} = \left(\cos \frac{2\pi}{5} + i \sin \frac{2\pi}{5}\right)^{5k} = \cos 2k\pi + i \sin 2k\pi = 1$$

Similarly $x^{2n} = (x^n)^2 = 1$, $x^{3n} = 1$, $x^{4n} = 1$

If n be a multiple of 5, then the sum of the nth powers of the roots $= 1 + x^n + x^{2n} + x^{3n} + x^{4n} = 1 + 1 + 1 + 1 + 1 = 5$. **Hence proved.**

Example 3:

If $\sin \alpha + \sin \beta + \sin \gamma = 0 = \cos \alpha + \cos \beta + \cos \gamma$, *show that*

(i) $\sin 2\alpha + \sin 2\beta + \sin 2\gamma = 0 = \cos 2\alpha + \cos 2\beta + \text{cog } 2\gamma$

(ii) $\sin^2 \alpha + \sin^2 \beta + \sin^2 \gamma = \dfrac{3}{2} = \cos^2 \alpha + \cos^2 \beta + \cos^2 \gamma$

Solution :

(i) Let $a = \cos \alpha + i \sin \alpha$, $b = \cos \beta + i \sin \beta$,

$c = \cos \gamma + i \sin \gamma$

Then $\dfrac{1}{a} = \dfrac{1}{\cos \alpha + i \sin \alpha} = (\cos \alpha + i \sin \alpha)^{-1} = \cos \alpha - i \sin \alpha$

Similarly $1/b = \cos \beta - i \sin \beta$ and $1/c = \cos \gamma - i \sin \gamma$

$\therefore \dfrac{1}{a} + \dfrac{1}{b} + \dfrac{1}{c} = (\cos \alpha - i \sin \alpha) + (\cos \beta - i \sin \beta) + (\cos \gamma - i \sin \gamma)$

$= (\cos \alpha + \cos \beta + \cos \gamma) - i (\sin \alpha + \sin \beta + \sin \gamma)$

$= (0) - i(0)$, $\quad \because \Sigma \sin \alpha = 0 = \Sigma \cos \alpha$

$= 0 \qquad \ldots(i)$

Contents

1

Introduction to De Moivre's Theorem

1.1 DE MOIVRE'S THEOREM

Statement : *Whatever be the value of n, positive or negative, integral or fractional,* $\cos n\theta + i \sin n\theta$ *is the value or one of the values of* $(\cos \theta + \sin \theta)^n$.

Proof :

Case 1: When n is a *positive integer.*

By actual multiplication we have

$(\cos \alpha + i \sin \alpha)(\cos \beta + \sin \beta)$

$= (\cos \alpha \cos \beta = \sin \alpha \sin \beta) + i(\cos \alpha \sin \beta = \sin \alpha \cos \beta)$

$= \cos(\alpha + \beta) + i \sin(\alpha + \beta)$

Similarly, $(\cos \alpha + i \sin \alpha)(\cos \beta + i \sin \beta)(\cos \gamma + i \sin \gamma)$

$= [\cos(\alpha + \beta) + i \sin(\alpha + \beta)](\cos \gamma + i \sin \gamma)$

$= [\cos(\alpha + \beta) \cos \gamma - \sin(\alpha + \beta) \sin \gamma]$

$+ i[\sin(\alpha + \beta) \cos \gamma + \cos(\alpha + \beta) \sin \gamma]$

$= \cos(\alpha + \beta + \gamma) + i(\alpha + \beta + \gamma)$

In this way we can prove that

$(\cos \alpha + i \sin \alpha)(\cos \beta + i \sin \beta)(\cos \gamma + i \sin \gamma)$

$(\cos \delta + i \sin \delta)$ ... to n factors

$= \cos(\alpha + \beta + \gamma + \delta + \text{... to n terms})$

$+ i \sin(\alpha + \beta + \gamma + \delta + \text{... to n terms})$

Putting $\alpha = \beta = \gamma = \delta = \ldots = \theta$, we have

$(\cos\theta + i\sin\theta)(\cos\theta + i\sin\theta)$ to n factors $= \cos n\theta + i\sin n\theta$

$\Rightarrow \quad (\cos\theta + i\sin\theta)^n = \cos n\theta + i\sin n\theta$

Case 2: When n is a negative integers, – m say). Where m itself is a positive integer.

In this case $(\cos\theta + i\sin\theta)^n = (\cos\theta + i\sin\theta)^{-m}$

$$= \frac{1}{(\cos\theta + i\sin\theta)^m}$$

$$= \frac{1}{\cos m\theta + i\sin m\theta} \quad \ldots \text{ by case I}$$

$$= \frac{1}{(\cos m\theta + i\sin m\theta)} \times \frac{(\cos m\theta - i\sin m\theta)}{(\cos m\theta + i\sin m\theta)},$$

multiplying both num. and denom. by $(\cos m\theta - i\sin m\theta)$

$$= \frac{\cos m\theta - i\sin m\theta}{\cos^2 m\theta - i^2\sin^2 m\theta}$$

$= \cos m\theta - i\sin m\theta \ldots \qquad \because i^2 = -1$

$= \cos(-m\theta) + i\sin(-m\theta)$

$= \cos n\theta + i\sin n\theta \ldots \qquad \because n = -m$

Case 3: When n is a fraction; say $n = \frac{p}{q}$, where θ is a positive integer and p is any integer positive or negative.

We have proved in cases I and II that

$(\cos\theta + i\sin\theta)^n = \cos n\theta + i\sin n\theta$, where n is any integer.

From here we can conclude that $\cos\theta + i\sin\theta$ is one of the values of $(\cos n\theta + i\sin n\theta)^{1/n}$, where n is any integer.

Now in this case 3.

$(\cos\theta + i\sin\theta)^n = (\cos\theta + i\sin\theta)^{p/q} = (\cos p\theta + i\sin p\theta)^{1/q}$ and as we have just shown above one of the values of

$$(\cos p\theta + i\sin p\theta)^{1/q} \text{ is } \cos\frac{p\theta}{q} + i\sin\frac{p\theta}{q}$$

Hence, De Moivre's Theorem is completly established

Note: $(\sin\theta + i\cos\theta)^n$ is not equal to $\sin n\theta + i\cos n\theta$.

1.2 GEOMETRICAL MEANING OF DE MOIVRE'S THEOREM

We know if $x + iy = r(\cos\theta + i\sin\theta)$, then it means (See Geometrical meaning of $x + iy$ in the chapter on complex numbers) a line of length r making an angle θ with OX or x-axis.

With the above meaning we can say that

$r[\cos(\theta + \phi) + i\sin(\theta + \phi)]$, means a line of the same length r making an angle $\theta + \phi$ with x-axis.

$\{r(\cos\theta + i\sin\theta)\}(\cos\phi + i\sin\phi) = r\{\cos(\theta + \phi) + i\sin(\theta + \phi)\}$

Hence, the effect of multiplying $r(\cos\theta + i\sin\theta)$ by $\cos\phi + i\sin\phi$ is to turn the line represented by $r(\cos\phi + i\sin\theta)$ or $x + iy$ through an angle f.

Hence the quantity

$(\cos\alpha + i\sin\alpha)(\cos\beta + i\sin\beta)(\cos\gamma + i\sin\gamma)(\cos\delta + i\sin\delta)$ means that the line represented by $(\cos\delta + i\sin\delta)$ is first turned through an angle γ, then through β and finally through a *i.e.*, altogether turned through an angle $\alpha + \beta + \gamma$ and this total operation as interpreted above gives the same line as

$$\{\cos(\alpha + \beta + \gamma) + i\sin(\alpha + \beta + \gamma)\}(\cos\delta + i\sin\delta)$$

A similar result can be derived for any number of factors.

Hence, by the De Moivre's Theorem the geometrical fact, that to turn a line through a number of angles successively has the same effect as turning through an angle equal to the sum of these angles, is expressed algebraically.

Example 1:

If $2\cos\theta = x + \frac{1}{x}$, $2\cos\phi = y + \frac{1}{y}$, *prove that*

(i) $2\cos(\theta + \phi) = xy + \frac{1}{xy}$ *(ii)* $2\cos(\theta - \phi) = \frac{x}{y} + \frac{y}{x}$

(iii) $x^m y^n + 1/(x^m y^n) = 2\cos(m\theta + n\phi)$.

(iv) $(x^m/y^n) + (y^n/x^m) = 2\cos(m\theta - n\theta)$.

Solution :

(i) $xy = (\cos\theta + i\sin\theta)(\cos\phi + i\sin\phi)$

$\Rightarrow xy = \cos(\theta + \phi) + i\sin(\theta + \phi)$ ...(1)

And $1/(xy) = [\cos(\theta + \phi) + i\sin(\theta + \phi)]^{-1}$

$\Rightarrow \quad 1/(xy) = \cos(\theta + \phi) - i\sin(\theta + \phi)$ ...(2)

Adding (1) and (2) we get $xy + 1/(xy) = 2\cos(\theta + \phi)$.

(ii) $x/y = (\cos\theta + i\sin\theta)(\cos\theta + i\sin\phi)^{-1}$ **(Note)**

$= (\cos\theta + i\sin\theta)[\cos(-\phi) + i\sin(-\phi)]$

$\Rightarrow \quad x/y = \cos(\theta - \phi) + i\sin(\theta - \phi)$ ...(3)

And $\quad y/x = 1/(x/y) = [\cos(\theta - \phi) + i\sin(\theta - \phi)]^{-1}$

$\Rightarrow \quad y/x = \cos(\theta - \phi) - i\sin(\theta - \phi)$ ...(4)

Adding (3) and (4) we get $(x/y) + (y/x) = 2\cos(\theta - \phi)$.

(iii) $x^m = (\cos\theta + i\sin\theta)^m = \cos m\theta + i\sin m\theta$

$x^n = (\cos\phi + i\sin\phi)^n = \cos n\phi + i\sin n\phi$

$\therefore \quad x^m y^n = (\cos m\theta + i\sin m\theta)(\cos n\phi + i\sin n\phi)$

$\Rightarrow \quad x^m y^n = \cos(m\theta + n\phi) + i\sin(m\theta + n\phi)$ ...(i)

Also $1/(x^m y^n) = [\cos(m\theta + n\phi) + i\sin(m\theta + n\phi)]^{-1}$

$\Rightarrow \quad 1/(x^m y^n) = \cos(m\theta + n\phi) - i\sin(m\theta + n\phi)$ ...(ii)

Adding (i) and (ii) we get

$x^m y^n + (1.x^m y^n) = 2\cos(m\theta + n\phi)$

(iv) $x^m/y^n = (\cos m\theta + i\sin m\theta)(\cos n\phi + i\sin n\phi)^{-1}$

$= (\cos m\theta + i\sin m\theta)[\cos(-n\phi) + i\sin(-n\phi)]$

$\Rightarrow \quad x^m/y^n = \cos(m\theta - n\phi) + i\sin(m\theta - n\phi)$ ...(iii)

Also $y^n/x^m = (x^m/y^n)^{-1} = [\cos(m\theta - n\phi) + i\sin(m\theta - n\phi)]^{-1}$

$\Rightarrow \quad y^n/x^m = \cos(m\theta - n\phi) - i\sin(m\theta - n\phi)$ ...(iv)

Adding (iii) and (iv) we get $(x^m/y^n) + (y^n/x^m) = 2\cos(m\theta - n\phi)$.

Example 2:

If a, b, c, d denote $\cos 2\alpha + i\sin 2\alpha$, $\cos 2\beta + i\sin 2\beta$, $\cos 2\gamma + i\sin 2\gamma$ and $\cos 2\delta + i\sin 2\delta$ respectively, show that

(i) $\sqrt{(abcd)} + \dfrac{1}{\sqrt{(abcd)}} = 2\cos(\alpha + \beta + \gamma + \delta)$;

(ii) $\sqrt{\left(\dfrac{ab}{cd}\right)} + \sqrt{\left(\dfrac{cd}{ab}\right)} = 2\cos(\alpha + \beta - \gamma - \delta)$

and (iii) $\sqrt{(a^p b^q c^r)} + \dfrac{1}{\sqrt{(a^p b^b c^r)}} = 2\cos(p\alpha + q\beta + r\gamma)$.

Solution :

(i) $abcd = (\cos 2\alpha + i \sin 2\alpha)(\cos 2\beta + i \sin 2\beta)$

$\times (\cos 2\gamma + i \sin 2\gamma)(\cos 2\delta + i \sin 2\delta)$

$= \cos(2\alpha + 2\beta + 2\gamma + 2\delta) + i \sin(2\alpha + 2\beta + 2\gamma + 2\delta)$

$\therefore \sqrt{(abcd)} = (abcd)^{1/2} [\cos(2\alpha + 2\beta + 2\gamma + 2\delta)$

$+ i \sin(2\alpha + 2\beta + 2\gamma + 2\delta)]^{1/2}$

$= \cos(\alpha + \beta + \gamma + \delta) + i \sin(\alpha + \beta + \gamma + \delta)$

and $\dfrac{1}{\sqrt{(abcd)}} = (abcd)^{-1/2} = [\cos(2\alpha + 2\beta + 2\gamma + 2\delta)$

$+ i \sin(2\alpha + 2\beta + 2\gamma + 2\delta)]^{-1/2}$

$= \cos(\alpha + \beta + \gamma + \delta) - i \sin(\alpha + \beta + \gamma + \delta)$

$\therefore \sqrt{(abcd)} + \dfrac{1}{\sqrt{(abcd)}}$

$= [\cos(\alpha + \beta + \gamma + \delta) + i \sin(\alpha + \beta + \gamma + \delta)]$

$+ (\cos(\alpha + \beta + \gamma + \delta) - i \sin(\alpha + \beta + \gamma + \delta)]$

$= \cos(\alpha + \beta + \gamma + \delta)$ **Hence proved.**

(ii) $\dfrac{ab}{cd} = \dfrac{(\cos 2\alpha + i \sin 2\alpha)(\cos 2\beta + i \sin 2\beta)}{(\cos 2\gamma + i \sin 2\gamma)(\cos 2\delta + i \sin 2\delta)}$

$= \dfrac{\cos(2\alpha + 2\beta) + i \sin(2\alpha + 2\beta)}{\cos(2\gamma + 2\delta) + i \sin(2\gamma + 2\delta)}$

$= [\cos(2\alpha + 2\beta) + i \sin(2\alpha + 2\beta)] [\cos(2\gamma + 2\delta) + i \sin(2\gamma + 2\delta)]^{-1}$

$= \cos(2\alpha + 2\beta - 2\gamma - 2\delta) + i \sin(2\alpha + 2\beta - 2\gamma - 2\delta)$

$\Rightarrow \sqrt{\left(\dfrac{ab}{cd}\right)} = \left(\dfrac{ab}{cd}\right)^{1/2} = [\cos 2(\alpha + \beta - \gamma - \delta) + i \sin 2(\alpha + \beta - \gamma - \delta)]^{-1/2}$

$= \cos(\alpha + \beta - \gamma - \delta) + i \sin(\alpha + \beta - \gamma - \delta)$

Similarly $\sqrt{\left(\dfrac{cd}{ab}\right)} = \left(\dfrac{ab}{cd}\right)^{-1/2}$

$= [\cos 2(\alpha + \beta - \gamma - \delta) + i \sin 2(\alpha + \beta - \gamma - \delta)]^{-1/2}$

$= \cos(\alpha + \beta - \gamma - \delta) - i \sin(\alpha + \beta - \gamma - \delta)$

$$\therefore \sqrt{\left(\frac{ab}{cd}\right)} + \sqrt{\left(\frac{cd}{ab}\right)} = [\cos(\alpha + \beta - \gamma - \delta) + i \sin(\alpha + \beta - \gamma - \delta)]$$
$$+ [\cos(\alpha + \beta - \gamma - \delta) - i \sin(\alpha + \beta - \gamma - \delta)]$$
$$= \cos(\alpha + \beta - \gamma - \delta)$$ **Hence proved.**

(iii) $a^p = (\cos 2\alpha + i \sin 2\alpha)^p = \cos 2p\alpha + i \sin 2p\alpha$

Similarly $b^q = \cos q\beta + i \sin 2q\beta$; $c^r = \cos 2r\gamma + i \sin 2r\gamma$

$$\therefore a^p b^q c^r = (\cos 2p\alpha + i \sin 2p\alpha)(\cos 2q\beta + i \sin 2q\beta)$$
$$\times (\cos 2r\gamma + i \sin 2r\gamma)$$
$$= \cos(2p\alpha + 2q\beta + 2r\gamma) + i \sin(2p\alpha + 2q\beta + 2r\gamma)$$
$$= \cos 2(p\alpha + q\beta + r\gamma) + i \sin 2(p\alpha + q\beta + r\gamma)$$
$$\therefore \sqrt{(a^p b^q c^r)} = [\cos 2(p\alpha + 2q\beta + r\gamma) + i \sin 2(p\alpha + q\beta + r\gamma)]^{1/2}$$
$$= \cos(p\alpha + q\beta + r\gamma) + i \sin(p\alpha + q\beta + r\gamma)$$

Similarly $\sqrt{\left(\frac{1}{a^p b^q c^r}\right)} = \cos(p\alpha + q\beta + r\gamma) - i \sin(p\alpha + q\beta + r\gamma)$

$$\therefore \sqrt{(a^p b^q c^r)} + \sqrt{\frac{1}{(a^p b^q c^r)}} = [\cos(p\alpha + q\beta + r\gamma) + i \sin(p\alpha + q\beta + r\gamma)]$$
$$+ [\cos(p\alpha + q\beta + r\gamma) - i \sin(p\alpha + q\beta + r\gamma)]$$
$$= 2\cos(p\alpha + qb + r\gamma)$$ **Hence proved.**

Example 3:

Show that

$$[(\cos\theta + \cos\phi) + i(\sin\theta + \sin\phi)]^n + [(\cos\theta + \cos\phi) - i(\sin\theta + \sin\phi)]^n$$
$$= 2^{n-1}\cos^n \frac{1}{2}(\theta - \phi) \cos \frac{1}{2} n(\theta + \phi).$$

Solution :

$$\text{L.H.S.} = [(\cos\theta + \cos\phi) + i(\sin\theta + \sin\phi)]^n + [(\cos\theta + \cos\phi) - i(\sin\theta + \sin\phi)]^n$$
$$= [(2\cos\frac{1}{2}(\theta + \phi)\cos\frac{1}{2}(\theta - \phi)\} + i\{2\sin\frac{1}{2}(\theta + \phi)\cos\frac{1}{2}(\theta - \phi)\}]^n$$
$$+ [\{2\cos\frac{1}{2}(\theta + \phi)\cos\frac{1}{2}(\theta - \phi)\} - i(2\sin\frac{1}{2}(\theta + \phi)\cos\frac{1}{2}(\theta - \phi)\}]^n$$

$$= [2 \cos \frac{1}{2} (\theta - \phi) \{\cos \frac{1}{2} (\theta + \phi) + i \sin \frac{1}{2} (\theta + \phi)]^n$$

$$+ [2 \cos \frac{1}{2} (\theta - \phi) \{\cos \frac{1}{2} (\theta + \phi) - i \sin \frac{1}{2} (\theta + \phi)\}]^n$$

$$= 2^n \cos^n \frac{1}{2} (\theta - \phi) [\cos \frac{1}{2} (\theta + \phi) + i \sin \frac{1}{2} (\theta + \phi)]^n$$

$$+ 2^n \cos^n \frac{1}{2} (\theta + \phi) [\cos \frac{1}{2} (\theta + \phi) - i \sin \frac{1}{2} (\theta + \phi)]^n$$

$$= 2^n \cos^n \frac{1}{2} (\theta - \phi) [\{\cos \frac{1}{2} n (\theta - \phi) + i \sin \frac{1}{2} n (\theta + \phi)\}$$

$$+ \{\cos \frac{1}{2} n (\theta - \phi) - i \sin \frac{1}{2} n (\theta + \phi)\}]$$

$$= 2^n \cos^n \frac{1}{2} (\theta - \phi) [2 \cos \frac{1}{2} n (\theta + \phi)]$$

$$= 2^{n+1} \cos^n \frac{1}{2} (\theta - \phi) \cos \frac{1}{2} n (\theta + \phi) = \text{R.H.S}$$

Example 4:

Show that

$$(1+\cos\theta + i \sin\theta)^n + (1+\cos\theta - i \sin\theta)^n = 2^{n \cdot 1} \cos^n \left(\frac{1}{2}\theta\right) .\cos \left(\frac{1}{2}n\theta\right).$$

Solution :

$$\text{L.H.S.} = (1 + \cos \theta + i \sin \theta)^n + (1 + \cos \theta - i \sin \theta)^n$$

$$= [2 \cos^2 \frac{1}{2}\theta + i \,.\, 2 \sin \frac{1}{2}\theta \cos \frac{1}{2}\theta]^n$$

$$+ [2 \cos^2 \frac{1}{2}\theta - i \,.\, 2 \sin \frac{1}{2}\theta \cos \frac{1}{2}\theta]^n$$

$$= 2^n \cos^n \left(\frac{1}{2}\theta\right) [\{\cos \frac{1}{2}\theta + i \sin \frac{1}{2}\theta \}^n + \{\cos \frac{1}{2}\theta - \sin \frac{1}{2}\theta)^n]$$

$$= 2^n \cos^n \left(\frac{1}{2}\theta\right) [(\cos \frac{1}{2}n\theta + i \sin \frac{1}{2}n\theta) + (\cos \frac{1}{2}n\theta - \sin \frac{1}{2}n\theta)]$$

$$= 2^n \cos^n \left(\frac{1}{2}\theta\right) [2 \cos \frac{1}{2} n\theta] = 2^{n+1} \cos^n \left(\frac{1}{2}\theta\right) \cos \left(\frac{1}{2}n\theta\right) = \text{R.H.S.}$$

Example 5:

Show that

$$\left(\frac{1+\cos\phi+i\sin\phi}{1+\cos\phi-i\sin\phi}\right)^n = \cos n\phi + i\sin n\phi.$$

Solution :

$$\text{L.H.S.} = \left[\frac{2\cos^2\frac{1}{2}\phi + i\,2\sin\frac{1}{2}\phi\cos\frac{1}{2}\phi}{2\cos^2\frac{1}{2}\phi - i.2\sin\frac{1}{2}\phi\cos\frac{1}{2}\phi}\right]^n$$

$$= \left[\frac{\cos\frac{1}{2}\phi + i\sin\frac{1}{2}\phi}{\cos\frac{1}{2}\phi - i\sin\frac{1}{2}\phi}\right]^n$$

$$= (\cos\tfrac{1}{2}\phi + i\sin\tfrac{1}{2}\phi)^n\,(\cos\tfrac{1}{2}\phi - i\sin\tfrac{1}{2}\phi)^{-n}$$

$$= (\cos\tfrac{1}{2}n\phi + i\sin\tfrac{1}{2}n\phi)\,(\cos\tfrac{1}{2}n\phi + i\sin\tfrac{1}{2}n\phi)$$

$$= (\cos\tfrac{1}{2}n\phi + i\sin\tfrac{1}{2}n\phi)^2$$

$$= (\cos n\phi + i\sin n\phi) = \text{R.H.S.}$$

Example 6:

Show that

$$\left[\frac{1+\sin\phi+i\cos\phi}{1+\sin\phi-i\cos\phi}\right]^n = \cos\left(\frac{1}{2}n\pi - n\phi\right) + i\sin\left(\frac{1}{2}n\pi - n\phi\right).$$

Solution :

$$\text{L.H.S.} = \left[\frac{1+\sin\phi+i\cos\phi}{1+\sin\phi\;\;i\cos\phi}\right]^n$$

$$= \left[\frac{1+\cos\theta+i\sin\theta}{1+\cos\theta-i\sin\theta}\right]^n, \text{ putting } \phi = \frac{1}{2}\pi - \theta$$

$$= \left[\frac{2\cos^2\frac{1}{2}\theta + i.2\sin\frac{1}{2}\theta\cos\frac{1}{2}\theta}{2\cos^2\frac{1}{2}\theta - i.2\sin\frac{1}{2}\theta\cos\frac{1}{2}\theta}\right]^n$$

$$\because 1 + \cos\theta = 2\cos^2\frac{1}{2}\theta \text{ and } \sin\theta = 2\sin\frac{1}{2}\theta\cos\frac{1}{2}\theta$$

$$= \left[\frac{2\cos\frac{1}{2}\theta\left(\cos\frac{1}{2}\theta + i\sin\frac{1}{2}\theta\right)}{2\cos\frac{1}{2}\theta\left(\cos\frac{1}{2}\theta - i\sin\frac{1}{2}\theta\right)}\right]^n = \left[\frac{\left(\cos\frac{1}{2}\theta + i\sin\frac{1}{2}\theta\right)}{\left(\cos\frac{1}{2}\theta - i\sin\frac{1}{2}\theta\right)}\right]^n$$

$$= [(\cos\frac{1}{2}\theta + i\sin\frac{1}{2}\theta) \times (\cos\frac{1}{2}\theta - i\sin\frac{1}{2}\theta)^{-1}]^n$$

$$= [(\cos\frac{1}{2}\theta + i\sin\frac{1}{2}\theta)(\cos\frac{1}{2}\theta + i\sin\frac{1}{2}\theta)]^n$$

$$= [(\cos\frac{1}{2}\theta + i\sin\frac{1}{2}\theta)^2]^n$$

$$= (\cos\theta + i\sin\theta)^n = \cos n\theta + i\sin n\theta$$

$$= \cos n\left(\frac{1}{2}\pi - \phi\right) + i\sin n\left(\frac{1}{2}\pi - \phi\right), \qquad \because \theta = \frac{1}{2}\pi - \theta$$

$$= \cos\left(\frac{1}{2}n\pi - n\phi\right) + i\sin\left(\frac{1}{2}n\pi - n\phi\right) = \text{R.H..S.}$$

Example 7:

Show that 4n the power of $(1 + 7i)/(2 - i)^2$ is equal to $(-4)^n$, where n is a positive integer.

Solution :

$$\frac{1+7i}{(2-i)^2} = \frac{(1+7i)(2+i)^2}{(2-i)^2(2+i)^2}$$, multiplying numerator and denominator by $(2 + i)^2$

$$= \frac{(1+7i)(4+4i+i^2)}{(4-i^2)^2} = \frac{(1+7i)(3+4i)}{(4+1)^2} \qquad \because i^2 = -1. \qquad ...(i)$$

$$= \frac{3+4i+21i+28i^2}{25} = \frac{3+25i-28}{25} = -1 + i.$$

Let $\quad -1 + i = r(\cos\theta + i\sin\theta)$. ...(ii)

Then $r\cos\theta = -1$, $r\sin\theta = 1$, equating real and imaginary parts on both sides

Squaring and adding these we get $r^2 = 2$ ...(iii)

Dividing we get $\tan\theta = -1$ which implies $\theta = -\frac{1}{4}\pi$, ...(iv)

taking the principal value of θ which lies between $-\pi$ and $+\pi$

$\therefore$ From (i) we get

$$\left[\frac{1+7i}{(2-i)^2}\right]^{4n} = [-1+i]^{4n}, \text{ from (i)}$$

$= [r(\cos\theta + i\sin\theta)]^{4n}$, from (ii)

$= (r)^{4n}(\cos\theta + i\sin\theta)^{4n}$

$= (r^2)^{2n}(\cos 4n\theta + i\sin 4n\theta)$ **(Note)**

$= (2)^{2n}\{\cos(-n\pi + i\sin n\pi\}$, where n is a positive integer.

$= 4^n\{(-1)^n + i(0)\}$, $\because \cos n\pi = (-1)^n$, $\sin n\pi = 0$.

$= 4^n(-1)^n = (-4)^n$. **Hence proved.**

Example 8:

Prove that

$(a+ib)^{m/n} + (a-ib)^{m/n} = 2(a^2+b^2)^{m/2n}\cos[(m/n)\tan^{1}(b/a)]$.

Solution :

Let $a = r\cos\theta$ and $b = r\sin\theta$

then squaring and adding we have $r^2 = a^2 + b^2 \Rightarrow r = \sqrt{(a^2+b^2)}$ ...(i)

and dividing we get $\tan\theta = b/a \Rightarrow \theta = \tan^{-1}(b/a)$ ...(ii)

$\therefore$ L.H.S. $= (a+ib)^{m/n} + (a-ib)^{m/n}$

$= [r\cos\theta + ir\sin\theta]^{m/n} + [r\cos\theta - ir\sin\theta]^{m/n}$,

putting $a = r\cos\theta$ and $b = r\sin\theta$

$= [r(\cos\theta + i\sin\theta)^{m/n} + [r(\cos\theta - i\sin\theta)]^{m/n}$

$= r^{m/n}(\cos\theta + i\sin\theta)^{m/n} + r^{m/n}(\cos\theta - i\sin\theta)^{m/n}$

$$= r^{m/n}\left[\left(\cos\frac{m\theta}{n} + i\sin\frac{m\theta}{n}\right) + \left(\cos\frac{m\theta}{n} - i\sin\frac{m\theta}{n}\right)\right]$$

$$= r^{m/n}\left[2\cos\frac{m\theta}{n} = 2r^{m/n}\cos\frac{m\theta}{n}\right]$$

$$= 2\left[\sqrt{(a^2+b^2)}\right]^{m/n} \cos\frac{m}{n}\left(\tan^{-1}\frac{b}{a}\right), \text{ from (i) and (ii)}$$

$$= 2\,(a^2+b^2)^{m/2n} \cos\frac{m}{n}\left(\tan^{-1}\frac{b}{a}\right) = \text{R.H.S.}$$

Example 9:

Show that $(1+i)^n + (1-i)^n = 2^{(n/2)-1} \cos\left(\frac{1}{4}n\pi\right)$.

Solution :

Let $1 + i = r(\cos\theta + i\sin\theta)$

Equating real and imaginary parts, we have

$r\cos\theta = 1$ and $r\sin\theta = 1$ ...(i)

Squaring and adding we have $r^2 = 1 + 1 = 2 \Rightarrow r = \sqrt{2}$

$\therefore$ From (i), $\cos\theta = 1/\sqrt{2}$ and $\sin\theta = 1/\sqrt{2}$, both of the equations are satisfied by $\theta = \frac{1}{4}\pi$

Hence $1 + i = \sqrt{2}\left(\cos\frac{1}{4}\pi + i\sin\frac{1}{4}\pi\right)$

Similarly $1 - i = \sqrt{2}\left(\cos\frac{1}{4}\pi - i\sin\frac{1}{4}\pi\right)$

Substituting these values of $(1+i)$ and $(1-i)$, we have

$$\text{L.H.S.} = \left[\sqrt{2}\left(\cos\frac{1}{4}\pi + i\sin\frac{1}{4}\pi\right)\right]^n + \left[\sqrt{2}\left(\cos\frac{1}{4}\pi - i\sin\frac{1}{4}\pi\right)\right]^n$$

$$= \left[2^{n/2}\left(\cos\frac{1}{4}n\pi + i\sin\frac{1}{4}n\pi\right)\right] - \left[2^{n/2}\left(\cos\frac{1}{4}n\pi - i\sin\frac{1}{4}n\pi\right)\right]$$

$$= 2^{n/2}\left[\left(\cos\frac{1}{4}n\pi + i\sin\frac{1}{4}n\pi\right) + \left(\cos\frac{1}{4}n\pi - i\sin\frac{1}{4}n\pi\right)\right]$$

$$= 2^{n/2}\left[2\cos\frac{1}{4}n\pi\right] = 2^{(n/2)+1}\cos\frac{1}{4}n\pi = \text{R.H.S.}$$

Example 10:

If $x_r = \cos(\pi/2^r) + i\sin(\pi/2^r)$, prove that $x_1 x_2 x_3$...ad. inf. = − 1.

Solution :

Given that $x_r = \cos(\pi/2^r) + i\sin(\pi/2^r)$

Putting r = 1, 2, 3, we have

$$x_1 = \cos\left(\frac{1}{2}\pi\right) + i\sin\left(\frac{1}{2}\pi\right),$$

$$x_2 = \cos\left(\frac{\pi}{2^2}\right) + i\sin\left(\frac{\pi}{2^2}\right),$$

$$x_3 = \cos\left(\frac{\pi}{2^3}\right) + i\sin\left(\frac{\pi}{2^3}\right), \text{...etc.}$$

$$x_2x_3x_3.. = \left[\cos\left(\frac{1}{2}\pi\right) + i\sin\left(\frac{1}{2}\pi\right)\right] \times$$

$$\left[\cos\left(\frac{\pi}{2^2}\right) + i\sin\left(\frac{\pi}{2^3}\right)\right] \times \left[\cos\left(\frac{\pi}{2^2}\right) + i\sin\left(\frac{\pi}{2^3}\right)\right] \times \text{...ad. inf.}$$

$$= \cos\left(\frac{1}{2}\pi + \frac{\pi}{2^2} + \frac{\pi}{2^3} + \text{...ad.inf.}\right) + i\sin\left(\frac{1}{2}\pi + \frac{\pi}{2^2} + \frac{\pi}{2^3} + \text{...ad.inf.}\right)$$

$$= \cos\left[\frac{1}{2}\pi\left(1 + \frac{1}{2} + \frac{1}{2^2} + \text{...ad.inf.}\right)\right] + i\sin\left[\frac{1}{2}\pi\left(1 + \frac{1}{2} + \frac{1}{2^2} + \text{...ad.inf.}\right)\right]$$

$$= \cos\left[\frac{\frac{1}{2}\pi}{1 - \frac{1}{2}}\right] + i\sin\left[\frac{\frac{1}{2}\pi}{1 - \frac{1}{2}}\right], \because 1 + r + r^2 + \text{... ad. inf.} = \frac{\text{"a"}}{1 - r}$$

$= \cos\pi + i\sin\pi = -1 - i(0) = -1$. **Hence proved.**

Example 11:

If $(a_1 + ib_1)(a_2 + ib_2)(a_3 + ib_3) \ldots (a_n + ib_n) = A + iB$, prove that

(i) $\left(a_1^2 + b_1^2\right)\left(a_2^2 + b_2^2\right)\left(a_3^2 + b_3^2\right) \ldots \left(a_n^2 + b_n^2\right) = A^2 + B^2$ and

(ii) $\tan^{-1}\frac{b_1}{a_1} + \tan^{-1}\frac{b_2}{a_2} + \tan^{-1}\frac{b_3}{a_3} + \ldots + \tan^{-1}\frac{b_n}{a_n} = \tan^{-1}\frac{B}{A}$

Solution :

Let $a_1 + ib_1 = r_1 (\cos \theta_1 + i \sin \theta_1)$;

$a_2 + ib_2 + r_2 (\cos \theta_2 + i \sin \theta_2)$,

and similar other expressions.

Then $r_1 = \sqrt{(a_1^2 + b_1^2)}$, $r_2 = \sqrt{(a_2^2 + b_2^2)}$ etc.

and $\theta_1 = \tan^{-1} \dfrac{b_1}{a_1}$, $\theta_2 = \tan^{-1} \dfrac{b_2}{a_2}$ etc. ...(a)

Now it is given that

$(a_1 + ib_1)(a_2 + ib_2)(a_3 + ib_3) ... (a_n + ib_n) = A + iB$

$\Rightarrow [r_1 (\cos \theta_1 + i \sin \theta_1)] [r_2 (\cos \theta_2 + i \sin \theta_2)]$

$... [r_n (\cos \theta_n + i \sin \theta_n)] = A + iB$,

substituting the values of $a_1 + ib_1$ etc.

$\Rightarrow r_1 r_2 r_3 ... r_n [(\cos \theta_1 + i \sin \theta_1)(\cos \theta_2 + i \sin \theta_2) ...$

$... (\cos \theta_n + i \sin \theta_n)] = A + iB$

$\Rightarrow r_1 r_2 r_3 \, r_n [\cos (\theta_1 + \theta_2 + ... + \theta_n) + i \sin (\theta_1 + \theta_2 + ... + \theta_n)] = A + iB$

Equating real and imaginary parts on both sides, we get

$A = r_1 r_2 r_3 ... r_n \cos (\theta_1 + \theta_2 + \theta_3 + ... + \theta_n)$ ...(b)

$B = r_1 r_2 r_3 ... r_n \sin (\theta_1 + \theta_2 + \theta_3 + ... + \theta_n)$ ...(c)

(i) Squaring and adding (b) and (c) we get

$A^2 + B^2 = r_1^2 r_2^2 r_3^3 ... r_n^2 = (a_1^2 + b_1^2)(a_2^2 + b_2^2) ... (a_n^2 + b_n^2)$ from (a)

(ii) Dividing (c) by (b) we get $B/A = \tan (\theta_1 + \theta_2 + ... + \theta_n)$

$\Rightarrow \tan^{-1} (B/A) = \theta_1 + \theta_2 + ... \theta_n$

$= \tan^{-1} \dfrac{b_1}{a_1} + \tan^{-1} \dfrac{b_2}{a_2} + \tan^{-1} \dfrac{b_3}{a_3} + ...$, from (a)

Example 12:

If $\left(1 + i\dfrac{x}{a}\right)\left(1 + i\dfrac{x}{b}\right)\left(1 + i\dfrac{x}{c}\right) ...\ A + iB$ *then prove that*

(i) $\left(1 + \dfrac{x^2}{a^2}\right)\left(1 + \dfrac{x^2}{b^2}\right)\left(1 + \dfrac{x^2}{c^2}\right) ... A^2 + B^2$

(ii) $\tan^{-1} \dfrac{x}{a} + \tan^{-1} \dfrac{x}{b} + \tan^{-1} \dfrac{x}{c} + ... = \tan^{-1} \dfrac{B}{A}$.

Solution :

Let $\frac{x}{a} = \tan\alpha,\ \frac{x}{b} = \tan\beta,\ \frac{x}{c} = \tan\gamma$, ...etc. ...(a)

Then the give result becomes

$$(1 + i\tan\alpha)(1 + i\tan\beta)(1 + i\tan\gamma)... = A + iB$$

$$\Rightarrow \left(1+i\frac{\sin\alpha}{\cos\alpha}\right)\left(1+i\frac{\sin\beta}{\cos\beta}\right)\left(1+i\frac{\sin\gamma}{\cos\gamma}\right)... = A + iB$$

$$\Rightarrow \left(\frac{\cos\alpha + i\sin\alpha}{\cos\alpha}\right)\left(\frac{\cos\beta + i\sin\beta}{\cos\beta}\right)\left(\frac{\cos\gamma + i\sin\gamma}{\cos\gamma}\right)... = A + iB$$

$$\Rightarrow \frac{1}{\cos\alpha\cos\beta\cos\gamma...}(\cos\alpha + i\sin\alpha)(\cos\beta + i\sin\beta)$$

$\times (\cos\gamma + i\sin\gamma)\ A + iB$

$\times \sec\alpha\sec\beta\sec\gamma\ ...[\cos(\alpha+\beta+\gamma+...) + i\sin(\alpha+\beta+\gamma+...)]$

$= A + iB$

Equating real and imaginary parts on both sides, we get

$A = \sec\alpha\sec\beta\sec\gamma\ ...\cos(\alpha+\beta+\gamma+...)$ ...(b)

$B = \sec\alpha\sec\beta\sec\gamma\ ...\sin(\alpha+\beta+\gamma+...)$ ...(c)

(i) Squaring and adding (b) and (c) we get

$A^2 + B^2 = \sec^2\alpha\ \sec^2\beta,\ \sec^2\gamma...,$

$= (1 + \tan^2\alpha)(1 + \tan^2\beta)(1 + \tan^2\beta)(1 + \tan^2\gamma)...$

$= \left(1+\frac{x^2}{a^2}\right)\left(1+\frac{x^2}{b^2}\right)\left(1+\frac{x^2}{c^2}\right)...$, from (a)

(ii) Dividing (c) by (b) we get $(B/A) = \tan(\alpha+\beta+\gamma+...)$

$\Rightarrow$ $\tan^{-1}(B/A) = \alpha + \beta + \gamma + ...$

$= \tan^{-1}\frac{x}{a} + \tan^{-1}\frac{x}{b} + \tan^{-1}\frac{x}{c} + ...$, from (a)

Hence proved.

Example 13:

$\cos\alpha + \cos\beta + \cos\gamma = \sin\alpha + \sin\beta + \sin\gamma = 0$, prove that $\cos 3\alpha + \cos 3\beta + \cos 3\gamma = 3\cos(\alpha+\beta+\gamma)$ and $\sin 3\alpha + \sin 3\beta + \sin 3\gamma = 3\sin(\alpha+\beta+\gamma)$.

Solution :

We know that if $a + b + c = 0$, $a^3 + b^3 + c^3 - 3abc = 0$

$\Rightarrow \quad a^3 + b^3 + c^3 = 3abc$ ...(i)

Now let $\alpha \cos \alpha + i \sin \alpha : b = \cos \beta + i \sin \beta$ and

$c = \cos \gamma + i \sin \gamma$

Then $a + b + c = (\cos \alpha + i \sin \alpha) + (\cos \beta + i \sin \beta) + (\cos \gamma + i \sin \gamma)$.

$- (\cos \alpha + \cos \beta + \cos \gamma) + i (\sin \alpha + \sin \beta + \sin \gamma)$

$\Rightarrow a + b + c = (0) + i (0)$ ...(given)

$= 0$

$\therefore$ From (i) we have $a^3 + b^3 + c^3 = 3abc$

$\Rightarrow (\cos \alpha + i \sin \alpha)^3 + (\cos \beta + i \sin \beta)^3 + (\cos \gamma + i \sin \gamma)^3$

$= 3 (\cos \alpha + i \sin \alpha) (\cos \beta + i \sin \beta) (\cos \gamma + i \sin \gamma)$

$\Rightarrow (\cos 3\alpha + i \sin 3\alpha) + (\cos 3\beta + i \sin 3\beta) + (\cos 3\gamma + i \sin 3\gamma)$

$= 3 [\cos (\alpha + \beta + \gamma) + i \sin (\alpha + \beta + \gamma)]$

$\Rightarrow (\cos 3\alpha + \cos 3\beta + \cos 3\gamma) + i (\sin 3\alpha + \sin 3\beta + \sin 3\gamma)$

$= 3 [\cos (\alpha + \beta + \gamma) + i \sin (\alpha + \beta + \gamma)]$

Equating real and imaginary parts on both sides we get

$\cos 3\alpha + \cos 3\beta + \cos 3\gamma = 3 \cos (\alpha + \beta + \gamma)$

and $\quad \sin 3\alpha + \sin 3\beta + \sin 3\gamma = 3 \sin (\alpha + \beta + \gamma)$. **Hence proved.**

Example 14:

Find the general value of q with satisfies the equations

$(\cos \theta + i \sin q) (\cos 2\theta + i \sin 2\theta)...(\cos r\theta + i \sin r\theta) = 1$

Solution :

$(\cos \theta + i \sin \theta) (\cos 2\theta + i \sin 2\theta) (\cos 3\theta + i \sin 3\theta)..$

$\times (\cos r\theta + i \sin r\theta) = 1$

$\Rightarrow \cos (\theta + r\theta + 3\theta + ... + r\theta) + i \sin (\theta + 2\theta + ... + r\theta) = 1$

$\Rightarrow \cos (1 + 2 + 3 + ... + r) \theta + i \sin (1 + 2 + 3 + ... + r) \theta = 1$

$$\Rightarrow \cos \left[\frac{1}{2} r (r+1) \theta\right] + i \sin \left[\frac{1}{2} r (r+1) \theta\right] = 1$$

$\because 1 + 2 + 3 + ... + r = \frac{1}{2} r (r + 1)$

Equating real parts on both sides, we get

$$\cos\left[\frac{1}{2}r\,(r+1)\,\theta\right] = 1 = \cos 0,$$

$$\therefore \frac{1}{2}r\ (r + 1)\ \theta = 2n\pi \pm 0, \text{ when n is zero or any integer,}$$

$$\Rightarrow\quad r\ (r + 1)\ \theta = 4n\pi$$

$$\Rightarrow\quad \theta = \frac{4n\pi}{r\,(r+1)}, \text{ where n is zero or any integer.}$$ **Ans.**

Example 15:

Find the general value of θ which satisfies the equation

$$(\cos\theta + i\sin\theta)\,(\cos 3\theta + i\sin 3\theta)\ldots\{\cos(2r-1)\,\theta + i\sin(2r\ \ 1)\,\theta\} = 1.$$

Solution :

$$(\cos\theta + i\sin\theta)\,(\cos 3\theta + i\sin 3\theta)\ldots$$
$$\ldots\{\cos(2r-1)\,\theta + i\sin(2r-1\theta) = 1$$

$$\Rightarrow \cos\{\theta + 3\theta + 5\theta + \ldots + (2r-1)\,\theta\} + i\sin\{\theta + 3\theta + 5\theta + \ldots + (2r-1)\,\theta\} = 1$$

$$\Rightarrow \cos\{1 + 3 + 5 + \ldots + (2r-1)\}\,\theta + i\sin\{1 + 3 + 5 + \ldots + (2r-1)\}\,\theta = 1$$

$$\Rightarrow \cos\left[\frac{1}{2}r\,\{2 + (r-1)\,2\}\right]\theta + i\sin\left[\frac{1}{2}r\,\{2 + (r-1)\,2\}\right]\theta = 1$$

$$\Rightarrow \cos(r^2\,\theta) + i\sin(r^2\,\theta) = 1$$

Equating real parts on both sides, we get

$$\cos(r^2\,\theta) = 1 = \cos 0$$

$\therefore r^2\,\theta\ 2n\pi \pm 0$, where n is zero or any integer.

$\Rightarrow \theta = 2n\pi/r^2$, where n is zero or any integer. **Ans.**

Example 16:

Find the equations whose roots are the nth powers of the roots of the equation $x^2 - 2x\cos\theta + 1 = 0$.

Solution :

The equation is $x^2 - 2x\cos\theta + 1 = 0$

$$\therefore x = \frac{2\cos\theta + \sqrt{(4\cos^2\theta - 4)}}{2} = \cos\theta \pm \sqrt{(\cos^2\theta - 1)}$$

$$= \cos\theta \pm \sqrt{(-\sin^2\theta)} = \cos\theta \pm i\sin\theta$$

$\therefore$ The roots of the given equation are $\cos\theta + i\sin\theta$ and $\cos\theta - i\sin\theta$. Hence the roots of the required equation are

$$(\cos\theta + i\sin\theta)^n \text{ and } (\cos\theta - i\sin\theta)^n$$

i.e., $(\cos n\theta + i\sin n\theta)$ and $(\cos n\theta - i\sin n\theta)$

$\therefore$ The sum of the roots of the required equation

$= (\cos n\theta + i\sin n\theta) + (\cos n\theta - i\sin n\theta) = 2\cos n\theta$

and the product of the required equation

$= (\cos n\theta - i\sin n\theta) \times (\cos n\theta - i\sin n\theta)$

$= \cos^2 n\theta - i2.\ \sin2\ n\theta$

$= \cos^2 n\theta + \sin^2 n\theta$ ... $\because i^2 = -1$

$= 1$

$\therefore$ The required equation is

$x^2 - x$ (sum of the roots) + (product of the roots) = 0

$\Rightarrow x^2 - 2x\cos n\theta + 1 = 0.$ **Ans.**

Example 17:

If $x = \cos\theta + i\sin\theta$ *and* $\sqrt{1 - c^2} = nc - 1$, *show that* $1 + c\cos\theta = \frac{c}{2n}(1 + nx)\left(1 + \frac{n}{x}\right)$.

Solution :

If $x = \cos\theta + i\sin\theta$,

then $\frac{1}{x} = (\cos\theta + i\sin\theta)^{-1} = \cos\theta - i\sin\theta$ ...(i)

Also $\sqrt{(1 - c^2)} = nc - 1$ or $1 - c^2 = (nc - 1)^2$, squaring both sides

$\Rightarrow \quad 1 - c^2 = n^2c^2 - 2nc + 1$

$\Rightarrow \quad c^2(n^2 + 1) = 2nc$

$\Rightarrow \quad c(n^2 + 1) = 2n$ ...(ii)

Now R.H.S. $= \frac{c}{2n}(1+nx)\left(1+\frac{n}{x}\right)$.

$= \frac{c}{2n}\left(1+nx+\frac{n}{x}+n^2\right) = \frac{c}{2n}(1+n^2) + \frac{c}{2n}n\left(x+\frac{1}{x}\right)$

$= \frac{c(1+n^2)}{2n} + \frac{c}{2}$ [(cos θ + i sin θ) + (cos θ – i sin θ)], from (1)

$= \frac{2n}{2n} + \frac{c}{2}$ [2 cos θ]... $\because c(1+n^2) = 2n$, from (ii)

= 1 + c cos θ = L.H.S. **Hence proved.**

Example 18:

Find the value of x such that

$$\frac{(x+\alpha)^n - (x+\beta)^n}{\alpha - \beta} = \frac{\sin n\theta}{\sin^n \theta'}$$ *where α and β are roots of the equation* $t^2 - 2t + 2 = 0$.

Solution :

The roots of the equation $t^2 - 2t + 2 = 0$ are given by

$$t = \frac{1}{2}\left[2 \pm \sqrt{(4-8)}\right] = \frac{1}{2}(2 \pm 2i) = 1 \pm i$$

Let α = 1 + i and β = 1 – i, as α and β are the roots of the given equation

∴ x + α = x + 1 + i = (x + 1) + i

and x + β = x + 1 – i = (x + 1) – i

Let (x + 1) + i = r (cos ϕ + i sin ϕ)

then r cos ϕ = x + 1 and r sin ϕ = 1

whence tan ϕ = 1/(x + 1) and $r = \sqrt{[1+(x+1)^2]} = \sqrt{[1+\cot^2\phi]}$,

$\because \cot\phi = (x+1)$

⇒r cosec ϕ

Hence x + α = r (cos ϕ + i sin ϕ) and x + β = r (cos ϕ – i sin ϕ),

where r = cosec ϕ and tan ϕ = 1/(x + 1)

$$\therefore \frac{(x+\alpha)^n - (x+\beta)^n}{\alpha - \beta} = \frac{\sin n\theta}{\sin^n \theta}$$

$$\Rightarrow \quad \frac{r^n (\cos\phi + i \sin\phi)^n - r^n (\cos\phi - i \sin\phi)^n}{(1+i)-(1-i)} = \frac{\sin n\theta}{\sin^n \theta}$$

$$\Rightarrow \quad \frac{r^n (2i \sin n\phi)}{2i} = \frac{\sin n\theta}{\sin^n \theta}$$

$$\Rightarrow \quad r^n \sin n\phi = \frac{\sin n\theta}{\sin^n \theta}$$

$$\Rightarrow \quad \frac{\sin n\phi}{\sin^n \phi} = \frac{\sin n\theta}{\sin^n \theta} \quad \ldots \quad \therefore r = \operatorname{cosec} \phi = \frac{1}{\sin \phi}$$

$$\Rightarrow \quad \phi = \theta \Rightarrow \tan\phi = \tan\theta$$

$$\Rightarrow \quad 1/(x+1) = \tan\theta \quad \ldots \quad \therefore \tan\phi = 1/(x-1)$$

$$\Rightarrow \quad x + 1 = \cot\theta$$

$$\Rightarrow \quad x = \cot\theta - 1.$$

Ans.

Example 19:

If α and β be the roots of $x^2 - 2x + 4 = 0$, prove that

$$\alpha^n + \beta^n = 2^{n-1} \cos \frac{1}{3} n\pi \quad \cos \frac{1}{3} n\pi.$$

Solution :

The roots of the equation $x^2 - 2x + 4 = 0$ are given by

$$x = \frac{2 \pm \sqrt{(4-1)}}{2} = 1 \pm 1\sqrt{3}$$

If α and β the roots of this equation, let $\alpha = 1 + \sqrt{3i}$ and $\beta = 1 - \sqrt{3i}$.

Let $1 + \sqrt{3i} = r(\cos\theta + i \sin\theta)$, then

$r \cos\theta = 1$ and $r \sin\theta = \sqrt{3}$ whence $r = 2$ and $\theta = \frac{1}{3}\pi$

$\therefore \alpha = r(\cos\theta + i \sin\theta)$ and $\beta = r(\cos\theta - \sin\theta)$

where $r = 2$ and $\theta = \frac{1}{3}\pi$.

$\therefore \alpha^n + \beta^n = r^n (\cos\theta + i \sin\theta)^n + r^n (\cos\theta - i \sin\theta)^n$

$\Rightarrow \alpha^n + \beta^n = r^n [(\cos n\theta + i \sin n\theta) + (\cos n\theta - i \sin n\theta)]$

$\Rightarrow \alpha^n + \beta^n = r^n (2 \cos n\theta) = 2^n \left[2 \cos \frac{1}{3} n\pi\right] \ldots \because r = 2$ and $\theta = \frac{1}{3}\pi$.

$= 2^{n+1} \cos \frac{1}{3} n\pi$

Hence proved.

Example 20:

Show that $(1+i)^n = 2^{n/2}\left[\cos\frac{1}{4}n\pi + i\sin\frac{1}{4}n\pi\right]$

Hence prove that if $(1+x)^n = p_0 + p_1x + p_2x^2 + p_3x^3 + \ldots$, *then*

(i) $p_0 - p_2 + p_4 \ldots = 2^{n/2}\sin\frac{1}{4}n\pi$

(ii) $p_1 - p_3 + p_5 - \ldots = 2^{n/2}\sin\frac{1}{4}n\pi$

Solution :

Let $1 + i = r(\cos\theta + i\sin\theta)$ then

$r\cos\theta = 1$ and $r\sin\theta = 1$

whence $r = \sqrt{(1+1)} = \sqrt{2}$ and $\tan\theta = 1 \Rightarrow \theta = \frac{1}{4}\pi$

$\therefore (1+i)^n = r^n(\cos\theta + i\sin\theta)^n = r^n(\cos n\theta + i\sin n\theta)$

$$= \left(\sqrt{2}\right)^n\left[\cos\frac{1}{4}n\pi + i\sin\frac{1}{4}n\pi\right]\ldots \qquad \because r = 2,\ \theta = \frac{1}{4}\pi$$

$$\therefore (1+i)^n = 2^{n/2}\left[\cos\frac{1}{4}n\pi + i\sin\frac{1}{4}n\pi\right] \qquad \ldots(a)$$

Again we are given that

$(1+x)^n = p_0 + p_1x + p_2x^2 + p_3x^3 + \ldots$

Putting $x = i$ on both sides we get

$(1+i)^n = p_0 + p_1i + p_2i^2 + p_3i^3 + p_4i^4 + \ldots$

$= p_0 + p_1i - p_2 - p_3i + p_4 + p_5i\ldots$

$= (p_0 - p + p_4 - p_6 + \ldots) + i(p_1 - p_3 + p_5 - \ldots)$

Substituting the value of $(1+i)^n$ from (a) we have,

$$2^{n/2}\left[\cos\frac{1}{4}n\pi + i\sin\frac{1}{4}n\pi\right] = (p_0 - p_2 + p_4 - \ldots) - i(p - p_3 + p_5 - \ldots)$$

Equating real and imaginary parts on both sides we have,

$$p_0 - p_2 + p_4 - \ldots = 2^{n/2}\cos\frac{1}{4}n\pi$$

and $$p_1 - p_3 + p_5 - \ldots = 2^{n/2}\sin\frac{1}{4}n\pi.$$

Example 21:

Evaluate *(a)* $(\cos\alpha - i\sin\alpha)^n$

(b) $(\cos\alpha + i\sin\alpha)^{-n}$

(c) $(\cos\alpha - i\sin\alpha)^{-n}$

Solution :

(a) $$(\cos\alpha - i\sin\alpha)^n = [\cos(-\alpha) + \sin(-\alpha)]^n$$
$$= \cos(-n\alpha) + i\sin(-n\alpha)$$
$$= \cos n\alpha - i\sin n\alpha \qquad \textbf{Ans.}$$

(b) $$(\cos\alpha + i\sin\alpha)^{-n} = \cos(-n\alpha) + \sin(-n\alpha)$$
$$= \cos n\alpha - i\sin n\alpha \qquad \textbf{Ans.}$$

(c) $$(\cos\alpha - i\sin\alpha)^{-n} = [\cos(-\alpha) + i\sin(-\alpha)]^{-n}$$
$$= \cos(n\alpha) + i\sin(n\alpha) \qquad \textbf{Ans.}$$

Example 22:

Simplify $\dfrac{(\cos\alpha + i\sin\alpha)^4}{(\cos\beta + i\sin\beta)^6}$.

Solution :

$$= \frac{(\cos\alpha + i\sin\alpha)^4}{(\cos\beta + i\sin\beta)^5} = \frac{\cos 4\alpha + i\sin 4\alpha}{\cos 5\beta + i\sin 5\beta}$$
$$= (\cos 4\alpha + i\sin 4\alpha)(\cos 5\beta + i\sin 5\beta)^{-1}$$
$$= (\cos 4\alpha + i\sin 4\alpha)[\cos(-5\beta) + i\sin(-5\beta)]$$
$$= \cos(4\alpha - 5\beta) + i\sin(4\alpha - 5\beta) \qquad \textbf{Ans.}$$

Example 23:

Simplify $\dfrac{(\cos\alpha + i\sin\alpha)^4}{(\sin\beta + i\cos\beta)^5}$.

Solution :

$$\frac{(\cos\alpha + i\sin\alpha)^4}{(\sin\beta + i\cos\beta)^5} = \frac{(\cos\alpha + i\sin\alpha)^4}{\left[\cos\left(\frac{1}{2}\pi - \beta\right) + i\sin\left(\frac{1}{2}\pi - \beta\right)\right]^5} \qquad \textbf{(Note)}$$

$$= \frac{\cos 4\alpha + i\sin 4\alpha}{\cos\left(\frac{5}{2}\pi - 5\beta\right) + i\sin\left(\frac{5}{2}\pi - 5\beta\right)} = \frac{\cos 4\alpha + i\sin 4\alpha}{\cos\left(\frac{1}{2}\pi - 5\beta\right) + i\sin\left(\frac{1}{2}\pi - 5\right.}$$

$\because \cos(2\pi+\theta)=\cos\theta,\ \sin(2\pi+\theta)=\sin\theta$

$$=(\cos 4\alpha + i\sin 4\alpha)\left[\cos\left(\frac{1}{2}\pi-5\beta\right)+i\sin\left(\frac{1}{2}\pi-5\beta\right)\right]^{-1}$$

$$=(\cos 4\alpha + i\sin 4\alpha)\left[\cos\left(-\frac{1}{2}\pi+5\beta\right)+i\sin\left(-\frac{1}{2}\pi+5\beta\right)\right]$$

$$=\cos\left(4\alpha-\frac{1}{2}\pi+5\beta\right)+i\sin\left(4\alpha-\frac{1}{2}\pi+5\beta\right)$$

$$=\cos\left[\frac{1}{2}\pi-(4\alpha+5\beta)\right]-i\sin\left[\frac{1}{2}\pi-(4\alpha+5\beta)\right]$$

$\because \cos(-\theta)=\cos\theta,\ \sin(-\theta)=-\sin\theta$

$=\sin(4\alpha+5\beta)-i\cos(4\alpha+5\beta)$ **Ans.**

Example 24:

Simplify $\dfrac{(\cos\theta+i\sin\theta)^4}{(\cos\theta-i\sin\theta)^3}$.

Solution :

$$\frac{(\cos\theta+i\sin\theta)^4}{(\cos\theta-i\sin\theta)^3}=\frac{\cos 4\theta+i\sin 4\theta}{\cos 3\theta-i\sin 3\theta}$$

$=(\cos 4\theta+i\sin 4\theta)(\cos 3\theta-i\sin 3\theta)^{-1}$

$=(\cos 4\theta+i\sin 4\theta)[\cos(-3\theta)+i\sin(-3\theta)]^{-1}$

$=(\cos 4\theta+i\sin 4\theta)(\cos 3\theta+i\sin 3\theta$

$=\cos 7\theta+i\sin 7\theta.$ **Ans.**

Example 25:

Simplify $\dfrac{(\cos 3\theta+i\sin 3\theta)^5(\cos\theta-i\sin\theta)^3}{(\cos 5\theta+i\sin 5\theta)^7(\cos 2\theta-i\sin 2\theta)^5}$.

Solution :

$$\frac{(\cos 3\theta+i\sin 3\theta)^5(\cos\theta-i\sin\theta)^3}{(\cos 5\theta+i\sin 5\theta)^7(\cos 2\theta-i\sin 2\theta)^5}$$

$$=\frac{(\cos 15\theta+i\sin 15\theta)[\cos(-\theta)+i\sin(-\theta)]^3}{(\cos 35\theta+i\sin 35\theta)[\cos(-2\theta)+i\sin(-2\theta)]^5}$$

$$= \frac{(\cos 15\theta + i \sin 15\theta)\left[\cos(-3\theta) + i \sin(-3\theta)\right]}{(\cos 35\theta + i \sin 35\theta)\left[\cos(-10\theta) + i \sin(-10\theta)\right]^5}$$

$$= \frac{\cos(15\theta - 3\theta) + i \sin(15\theta - 3\theta)}{\cos(35\theta - 10\theta) + i \sin(35\theta - 10\theta)}$$

$$= \frac{(\cos 12\theta + i \sin 12\theta)}{(\cos 25\theta + i \sin 25\theta)}$$

$= (\cos 12\theta + i \sin 12\theta)(\cos 25\theta + i \sin 25\theta)^{-1}$

$= \cos(12\theta - 12\theta)\,[\cos(-25\theta) + i \sin(-25\theta)]$

$= \cos(12\theta - 25\theta) + i \sin(12\theta - 25\theta)$

$= \cos(-13\theta) + i \sin(-13\theta) = (\cos 13\theta - i \sin 13\theta)$. **Ans.**

Example 26:

If $x = \cos\theta + i \sin\theta$; $y = \cos\phi + i \sin\phi$, *show that*

$$\frac{x-y}{x+y} = i \tan\frac{\theta-\phi}{2}.$$

Solution :

$$\text{L.H.S.} = \frac{x-y}{x+y} = \frac{(\cos\theta + i \sin\theta) - (\cos\phi + i \sin\phi)}{(\cos\theta + i \sin\theta) + (\cos\phi + i \sin\phi)}$$

$$= \frac{(\cos\theta - i \sin\phi) - i(\sin\theta - \sin\phi)}{(\cos\theta + i \sin\phi) + i(\sin\theta + \sin\phi)}$$

$$= \frac{\left[2 \sin\frac{1}{2}(\theta+\phi) \sin\frac{1}{2}(\phi-\theta)\right] + i\left[2 \cos\frac{1}{2}(\theta+\phi) \sin\frac{1}{2}(\theta-\phi)\right]}{\left[2 \cos\frac{1}{2}(\theta+\phi) \cos\frac{1}{2}(\theta-\phi)\right] + i\left(2 \sin\frac{1}{2}(\theta+\phi) \cos\frac{1}{2}(\theta-\phi)\right)}$$

$$= \frac{2 \sin\frac{1}{2}(\theta-\phi)\left[\sin\frac{1}{2}(\theta+\phi) + i \cos\frac{1}{2}(\theta+\phi)\right]}{2 \cos\frac{1}{2}(\theta-\phi)\left[\cos\frac{1}{2}(\theta+\phi) + i \sin\frac{1}{2}(\theta+\phi)\right]},$$

$$\because \sin\frac{1}{2}(\phi-\theta) = -\sin\frac{1}{2}(\theta-\phi)$$

$$= \frac{\sin\frac{1}{2}(\theta-\phi)\left[i\left\{i\sin\frac{1}{2}(\theta+\phi)+i\cos\frac{1}{2}(\theta+\phi)\right\}\right]}{\cos\frac{1}{2}(\theta-\phi)\left[\cos\frac{1}{2}(\theta+\phi)+i\sin\frac{1}{2}(\theta+\phi)\right]}, \quad \because i^2=-1$$

$$= \tan\frac{1}{2}(\theta-\phi)\,[i] = i\tan\frac{1}{2}(\theta-\phi) = \text{R.H.S.}$$ **Ans.**

Example 27:

If a denotes cos 2α + i sin 2a, with similar expressions for b, c, d prove that

(i) $ab + cd = 2\cos(\alpha+\beta+\gamma-\delta)\,[\cos(\alpha+\beta+\gamma+\delta) + i\sin(\alpha+\beta+\gamma+\delta)]$

(ii) $(a+b)(c+d) = 4\cos(\alpha-\beta)\cos(\gamma-\delta\,[\cos(\alpha+\beta+\gamma+\delta) + i\sin(\alpha+\beta+\gamma+\delta)]$

Solution :

Let $b = \cos 2\beta + i\sin 2\beta,\ c = \cos 2\gamma + i\sin 2\gamma$

and $d = \cos 2\delta + i\sin 2\delta$

(i) $ab = (\cos 2\alpha + i\sin 2\alpha)(\cos 2\beta + i\sin 2\beta)$

$= (\cos 2\alpha\cos 2\beta - \sin 2\alpha\sin 2\beta) + i(\sin 2\alpha\cos 2\beta + \cos 2\alpha\sin 2\beta)$

$= \cos(2\alpha+2\beta) + i\sin(2\alpha+2\beta)$

Similarly it can be proved that

$cd = \cos(2\gamma+2\delta) + i\sin(2\gamma+2\delta)$

$\therefore ab + cd = [\cos(2\alpha+2\beta) + i\sin(2\alpha+2\beta)] + [\cos(2\gamma+2\delta) + i\sin(2\gamma+2\delta)]$

$= [\cos(2\alpha+2\beta) + \cos(2\gamma+2\delta] + i[\sin(2\alpha+2\beta) + \sin(2\gamma+2\delta)]$

$= [2\cos(\alpha+\beta+\gamma+\delta)\cos(\alpha+\beta-\gamma-\delta)] + i[2\sin(\alpha+\beta+\gamma+\delta)\cos(\alpha+\beta-\gamma-\delta)]$

$= 2\cos(\alpha+\beta-\gamma-\delta)\,[\cos(\alpha+\beta+\gamma+\delta) + i\sin(\alpha+\beta+\gamma+\delta)]$

Hence proved.

(ii) $a + b = (\cos 2\alpha + i\sin 2\alpha) + (\cos 2\beta + i\sin 2\beta)$

$= (\cos 2\alpha + \cos 2\beta) + i(\sin 2\alpha + \sin 2\beta)$

$= 2 \cos(\alpha - \beta) \cos(\alpha + \beta) + i [2 \sin(\alpha + \beta) \cos(\alpha - \beta)]$

$= 2 \cos(\alpha - \beta) [\cos(\alpha + \beta) + i \sin(\alpha + \beta)]$.

Similarly we can prove that

$c + d = 2 \cos(\gamma - \delta) [\cos(\gamma + \delta) + i \sin(\gamma + \delta)]$

$\therefore (a + b)(c + d) = [2 \cos(\alpha - \beta) \{\cos(\alpha + \beta) + i \sin(\alpha + \beta)\}]$
$\times [2 \cos(\gamma - \delta) \{\cos(\gamma + \delta) + i \sin(\gamma + \delta)\}]$

$= 4 \cos(\alpha - \beta) \cos(\gamma + \delta) [\cos(\alpha + \beta) + i \sin(\alpha + \beta)]$
$\times [\cos(\gamma + \delta) + i \sin(\gamma + \delta)]$

$= \cos(\alpha - \beta) \cos(\gamma - \delta) [\{\cos(\alpha + \beta) \cos(\gamma + \delta) - \sin(\alpha + \beta)$
$\times \sin(\gamma + \delta)\} + i \{\sin(\alpha + \beta) \cos(\gamma + \delta) + \cos(\alpha + \beta) \sin(\gamma + \delta)\}]$

$= 4 \cos(\alpha - \beta) \cos(\gamma - \delta) [\cos(\alpha + \beta + \gamma + \delta) + i \sin(\alpha + \beta + \gamma + \delta)]$

Hence proved.

Example 28:

If $\cos\theta = a + \dfrac{1}{a}$ *:* $2 \cos\phi = b + \dfrac{1}{b}$*; etc. show that*

(i) $abc... + \dfrac{1}{abc....} = 2 \cos(\theta + \phi + ...)$

(ii) $a^p b^q c^r + ... + \dfrac{1}{a^p b^q c^r .....}$ $2 \cos(p\theta + q\phi + ...)$

Solution :

Given $a + (1/a) = 2 \cos\theta$

$\Rightarrow \quad a^2 + 1 = 2 \cos\theta$

$\Rightarrow \quad a^2 - 2\alpha \cos\theta + 1 = 0$

$$\Rightarrow \quad a = \frac{2 \cos\theta + \sqrt{(4 \cos^2\theta - 4)}}{2}$$

$$= \cos\theta + \sqrt{(\cos^2\theta - 1)} = \cos\theta + \sqrt{(-\sin^2\theta)}.$$

$\Rightarrow \quad a = \cos\theta + i \sin\theta. \qquad \because i^2 = -1$

Similarly $b = \cos\phi + i \sin\phi$, etc.

(i) abc. $= (\cos\theta + i \sin\theta)(\cos\phi + i \sin\phi)(\cos\phi + i \sin\phi)(\cos\psi + i \sin\psi)...$

$= \cos(\theta + \phi + \psi + ...) + i \sin(\theta + \phi + \psi + ...)$

and $\dfrac{1}{abc...} = (abc)...)^{-1}$

$= [\cos(\theta + \phi + \psi + ...) + i \sin(\theta + \phi + \psi + ...)]^{-1}$

$= \cos(\theta + \phi + \psi + ...) - i \sin(\theta + \phi + \psi + ...)$

$\therefore\ abc + ... \dfrac{1}{abc...} = [\cos(\theta + \phi + \psi + ...) + i \sin(\theta + \phi + \psi + ...)]$

$+ [\cos(\theta + \phi + \psi + ...) - i \sin(\theta + \phi + \psi + ...)]$

$= 2 \cos(\theta + \phi + \psi + ...)$

(ii) $a^p = (\cos\theta + i \sin\phi)^p = \cos p\theta + i \sin p\theta$;

$b^q = (\cos\phi + i \sin\phi)^q = \cos q\phi + i \sin q\phi$; ... etc.

$\therefore\ a^p b^q c^r ... = (\cos p\theta + i \sin p\theta)(\cos q\phi + i \sin q\phi)...$

$= \cos(p\theta + q\phi + ...) + i \sin(p\theta + q\phi + ...)$

And $1/a^p b^q c^r ... = (a^p b^q c^r ...)^{-1}$

$= [\cos(p\theta + q\phi + ...) + i \sin(p\theta + q\phi ...)]^{-1}$

$= \cos(p\theta + q\phi + ...) - i \sin(p\theta + q\phi + ...)$

$\therefore\ a^p b^q c^r ... + \dfrac{1}{a^p b^q c^r ...} = [\cos(p\theta + q\phi + ...) + i \sin(p\theta + q\phi + ...)]$

$+ [\cos(p\theta + q\phi + ...) - i \sin(p\theta + q\phi + ...)]$

$= 2 \cos(p\theta + q\phi + ...)$ **Hence proved.**

2

To Find the Q Roots of (Cos θ – in Sin θ)$^{P/Q}$, Where P and Q are Integers Prime to Each Other

In we have shown that $\cos \dfrac{p\theta}{q} + i \sin \dfrac{p\theta}{q}$ is one of the values of $(\cos \theta + \sin \theta)^{p/q}$. Here in this article we shall find out all the q different values of $(\cos \theta + i \sin \theta)^{p/q}$.

$\therefore$ We know that, $\cos (2n\pi + \theta) = \cos \theta$ and $\sin (2n\pi + \theta) = \sin \theta$, where n is any integer.

$$\therefore \cos \theta + i \sin \theta = \cos (2n\pi + \theta) + i \sin (2n\pi + \theta)$$

$$\Rightarrow (\cos \theta + i \sin \theta)^{p/q} = [\cos (2n\pi + \theta) + i \sin (2n\pi + \theta)]^{p/q}$$

$$= \cos \left\{\frac{p\,(2n\pi + \theta)}{q}\right\} + i \sin \left\{\frac{p\,(2n\pi + \theta)}{q}\right\}$$

Giving n the successive values of 0, 1, 2, 3,..., (q –1) we obtain the q values as

$$\cos \frac{p\theta}{q} + i \sin \frac{p\theta}{q}, \qquad \text{when } n = 0$$

$$\cos \frac{p\,(2\pi + \theta)}{q} + i \sin \frac{p\,(2\pi + \theta)}{q} \qquad \text{when } n = 1$$

$$\cos \frac{p\,(4\pi + \theta)}{q} + i \sin \frac{p\,(4\pi + \theta)}{q} \qquad \text{when } n = 2$$

$$\cos \frac{p}{q} \{2\,(\theta - 1)\,\pi + \theta\} + i \sin \frac{p}{q} \{2\,(q - 1)\,\pi + \theta\}, \text{ when } n = q-1$$

and these are the q different values of $(\cos \theta + i \sin \theta)^{p/q}$

Note: The highest value that should be given to n is (q – 1).

If we give n the integral values greater than (q – 1) then we shall not get any new values but same values as given by

$$n = 0, 1, 2, 3 ....$$

For example let n = q, then we get the corresponding value as

$$\cos \frac{p\,(2q\pi + \theta)}{q} + i \sin \frac{p\,(2q\pi + \theta)}{q}.$$

$$= \cos p \left(2\pi + \frac{6}{q}\right) + i \sin p\left(2\pi + \frac{\theta}{q}\right)$$

$$= \cos \left(2p\pi + \frac{p\theta}{q}\right) + i \sin \left(2p\pi + \frac{p\theta}{q}\right)$$

$$= \cos \frac{p\theta}{q} + i \sin \frac{p\theta}{q}, ... \qquad \because \begin{cases} \cos (2p\pi + \theta) = \cos \theta \\ \sin (2p\pi + \theta) = \sin \theta \end{cases}$$

which is the same value as obtained by putting n = 0.

Example 1:

Find the cube roots of (1 – cos ϕ – i sin ϕ), where ϕ i real.

Solution :

Let 1 – cos ϕ – i sin ϕ = r (cos ϕ – i sin ϕ)

Equating real and imaginary parts we have

$$r \cos \theta = 1 - \cos \phi \text{ and } r \sin \theta = \sin \phi$$

$$\text{whence } \quad r = \sqrt{\left\{(1 - \text{co } \phi)^2 + \sin^2 \phi\right\}} = \sqrt{(2 - 2 \cos \phi)}$$

$$= \sqrt{\{2\,(1 - \cos \phi)\}} = \sqrt{\left\{2\left(2 \sin^2 \frac{1}{2}\phi\right)\right\}} = 2 \sin \frac{1}{2}\phi \text{ and}$$

$$\tan \phi = \frac{\sin \phi}{1 - \cos \phi} = \frac{2 \sin \frac{1}{2}\phi \cos \frac{1}{2}\phi}{2 \sin^2 \frac{1}{2}\phi} = \cot \frac{1}{2}\phi = \tan \left(\frac{1}{2}\pi - \frac{1}{2}\phi\right)$$

$$\Rightarrow \qquad \theta = \frac{1}{2}\pi - \frac{1}{2}\phi \qquad \text{...(ii)}$$

Now $(1 - \cos \phi - i \sin \phi)^{1/3} = r^{1/3} (\cos \phi - i \sin \theta)^{1/3}$

$$= r^{1/3} [\cos (2n\pi + \theta) - i \sin (2n\pi + \theta)]^{1/3}$$

$$= r^{1/3}\left[\cos\left(\frac{2n\pi+\theta}{3}\right) - i\sin\left(\frac{2n\pi+\theta}{3}\right)\right]$$

$$= \left(2\sin\frac{\phi}{2}\right)^{1/3}\left[\cos\left\{\frac{2n\pi+\frac{1}{2}\pi-\frac{1}{2}\phi}{3}\right\} - \sin\left(\frac{2n\pi+\frac{1}{2}\pi-\frac{1}{2}\phi}{3}\right)\right]$$

putting the values of r and θ from (i) and (ii)

$$= \left(2\sin\frac{\phi}{2}\right)^{1/3}\left[\cos\left\{\frac{4n\pi+\pi-\phi}{6}\right\} - \sin\left(\frac{4n\pi+\pi-\phi}{6}\right)\right]$$

∴ The required cube roots of $(1 - \cos\phi - i\sin\phi)$ are given by

$$\left(2\sin\frac{\phi}{2}\right)^{1/3}\left[\cos\left\{\frac{4n\pi+\pi-\phi}{6}\right\} - \sin\left(\frac{4n\pi+\pi-\phi}{6}\right)\right]$$

where n = 0, 1, 2.

Example 2:

Find the five fifth roots of unity (or solve the equation $x^5 = 1$) and prove that sum of their nth power always vanishes unless n be a multiple of 5, n being an integer and that then the sum is 5.

Solution :

We know $\quad 1 = \cos 2n\pi + i\sin 2n\pi$

then $(1)^{1/5} = (\cos 2n\pi + i\sin 2n\pi)^{1/5} = \cos\dfrac{2n\pi}{5} + i\sin\dfrac{2n\pi}{5}$

Giving n the values 0, 1, 2, 3 and 4 we have the five 5th roots of unity

as $(\cos 0 + i\sin 0)$, $\left(\cos\dfrac{2\pi}{5} + i\sin\dfrac{2\pi}{5}\right)$,

$\left(\cos\dfrac{4\pi}{5} + i\sin\dfrac{4\pi}{5}\right)$, $\left(\cos\dfrac{6\pi}{5} + i\sin\dfrac{6\pi}{5}\right)$ and $\left(\cos\dfrac{8\pi}{5} + i\sin\dfrac{8\pi}{5}\right)$

i.e., 1, x, x^2, x^3 and x^5,

where x = cos $\cos\dfrac{2\pi}{5} + i\sin\dfrac{2\pi}{5}$.

Sum of the nth powers of these roots

$= 1 + x^n + x^{2n} + x^{3n} + x^{4n}$.

$$= \frac{1-(x^n)^5}{1-x^n} \ldots \quad \therefore\ 1 + r + r^2 + \ldots\text{to n terms} = \frac{1-r^n}{1-r}$$

Now $x = \cos\ \cos\dfrac{2\pi}{5} + i\sin\dfrac{2\pi}{5}$, $\because\ x^{5n} = \left(\cos\dfrac{2\pi}{5} + i\sin\dfrac{2\pi}{5}\right)^{5n}$

$\Rightarrow x^{5n} = \cos 2n\pi + i\sin 2n\pi = 1$

Also $x^n = \left(\cos\dfrac{2\pi}{5} + i\sin\dfrac{2\pi}{5}\right)^n = \cos\dfrac{2n\pi}{5} + i\sin\dfrac{2n\pi}{5}$

which is equal to 1, only if n be a multiple of 5, otherwise not.

$\therefore$ If n is not a multiple of 5, x^n is not equal to one and in this case the sum of the nth powers of the roots

$$= \frac{1-x^{5n}}{1-x^n} = \frac{0}{1-x^n} = 0, \text{ as } 1 - x^n \neq 0$$

Again if n is a multiple of 5, then n = 5k (say)

$$\therefore x^n = x^{5k} = \left(\cos\frac{2\pi}{5} + i\sin\frac{2\pi}{5}\right)^{5k} = \cos 2k\pi + i\sin 2k\pi = 1$$

Similarly $x^{2n} = (x^n)^2 = 1$, $x^{3n} = 1 = x^{4n}$

$\therefore$ If n be a multiple of 5, then the sum of the nth powers of the roots $= 1 + x^n + x^{2n} + x^{3n} + x^{3n} + x^{4n} = 1 + 1 + 1 + 1 + 1 = 5$. **Hence proved.**

Example 3:

If $\sin\alpha + \sin\beta + \sin\gamma = 0\ \cos\alpha + \cos\beta + \cos\gamma$, show that

(i) $\sin 2\alpha + \sin 2\beta + \sin 2\gamma = 0 = \cos 2\alpha + \cos 2\beta + cog\ 2\gamma$

(ii) $\sin^2\alpha + \sin^2\beta + \sin^2\gamma = \dfrac{3}{2} = \cos^2\alpha + \cos^2\beta + \cos^2\gamma$.

Solution :

(i) Let $a = \cos\alpha + i\sin\alpha$, $b = \cos\beta + i\sin\beta$,

$c = \cos\gamma + i\sin\gamma$

Then $\dfrac{1}{\alpha} = \dfrac{1}{\cos\alpha + i\sin\alpha} = (\cos\alpha + i\sin\alpha)^{-1} = \cos\alpha - i\sin\alpha$

Similarly $1/b = \cos\beta - i\sin\beta$ and $1/c = \cos\gamma - i\sin\gamma$

$\therefore\ \dfrac{1}{a} + \dfrac{1}{b} + \dfrac{1}{c} = (\cos\alpha - i\sin\alpha) + (\cos\beta - i\sin\beta) + (\cos\gamma - i\sin\gamma)$

$= (\cos\alpha + \cos\beta + \cos\gamma) - i(\sin\alpha + \sin\beta + \sin\gamma)$

$= (0) - i(0)$, $\qquad \because \Sigma\sin\alpha = 0 = S\cos\alpha$

$= 0 \qquad$ ...(i)

Similarly $a + b + c = 0$ ...(ii)

(i) Now $(a + b + c)^2 = (a^2 + b^2 + c^2) + 2(ab + bc + ca)$

$$= (a^2 + b^2 + c^2) + 2abc\,(1/c + 1/a + 1/b)$$

$\Rightarrow 0 = (a^2 + b^2 + c^2) + 2abc\,(0)$, from (i) and (ii)

$\Rightarrow \quad a^2 + b^2 + c^2 = 0$

$\Rightarrow (\cos\alpha + i\sin\alpha)^2 + (\cos\beta + i\sin\beta)^2 + (\cos\gamma + i\sin\gamma)^2 = 0$

$\Rightarrow (\cos 2\alpha + i\sin 2\alpha) + (\cos 2\beta + i\sin 2\beta) + (\cos 2\gamma + i\sin 2\gamma) = 0$

$\Rightarrow (\cos 2\alpha + \cos 2\beta + \cos 2\gamma) + i(\sin 2\alpha + \sin 2\beta + \sin 2\gamma) = 0$

Equating real and imaginary parts on both sides, we get

$\cos 2\alpha + \cos 2\beta + \cos 2\gamma = 0 = \sin 2\alpha + \sin 2\beta + \sin 2\gamma$

Hence proved.

(ii) If $\cos 2\alpha + \cos 2\beta + \cos 2\gamma = 0$

then $(2\cos^2\alpha - 1) + (2\cos^2\beta - 1) + (2\cos^2\gamma - 1) = 0$

$\Rightarrow \quad 2(\cos^2\alpha + \cos^2\beta + \cos^2\gamma) = 3$

$\Rightarrow \quad \cos^2\alpha + \cos^2\beta + \cos^2\gamma = \dfrac{3}{2}$ **Hence proved.**

Again $\quad \cos 2\alpha + \cos 2\beta + \cos 2\gamma = 0$

$\Rightarrow \quad (1 - 2\sin^2\alpha) + (1 - 2\sin^2\beta) + (1 - 2\sin^2\gamma) = 0$

$\Rightarrow \quad 3 = 2(\sin^2\alpha + \sin^2\beta + \sin^2\gamma).$

$\Rightarrow \quad \sin^2\alpha + \sin^2\beta + \sin^2\gamma = \dfrac{3}{2}.$ **Hence proved.**

Example 4:

Express $2 - \sqrt{3} + i$ *in the form* $r(\cos\theta + i\sin\theta)$.

Solution :

Let $\left(2 - \sqrt{3}\right) + i = r(\cos\theta + i\sin\theta)$ ...(i)

Since $x \neq 0$ so the given equation can be written as

$$\left(\frac{x-1}{x}\right)^n = 1 = \cos 2r\pi + i\sin 2r\pi \qquad \textbf{(Note)}$$

$$\Rightarrow \left(1 - \frac{1}{x}\right) = (\cos 2r\pi + i\sin 2r\pi)^{1/n}$$

$$\Rightarrow 1 - \frac{1}{x} = \cos\left(\frac{2r\pi}{n}\right) + i\sin\left(\frac{2r\pi}{n}\right), \text{ where } r = 0, 1, 2, \ldots, (n-1)$$

$$\Rightarrow \frac{1}{x} = \left\{1 - \cos\left(\frac{2r\pi}{n}\right)\right\} - i\left\{\sin\left(\frac{2r\pi}{n}\right)\right\}$$ **(Note)**

$= 2 \sin^2 (r\pi/n) - i \{2 \sin (r\pi/n) \cos (r\pi/n)\}$

$= 2 \sin (r\pi/n) \{\sin (r\pi/n) - i \cos (r\pi/n)\}$

$$= \frac{2 \sin (r\pi/n) \{\sin^2 (r\pi/n) - i^2 \cos^2 (r\pi/n)\}}{\sin (r\pi/n) + i \cos (r\pi/n)},$$ **(Note)**

multiplying num. and denom. by $\sin\left(\frac{r\pi}{n}\right) + i \cos\left(\frac{r\pi}{n}\right)$

$$\Rightarrow \frac{1}{x} = \frac{2 \sin (r\pi/n) \{1\}}{\sin (r\pi/n) + i \cos (r\pi/n)}, \qquad \because i^2 = -1$$

$$\Rightarrow x = \frac{\sin (r\pi/n) + i \cos (r\pi/n)}{2 \sin (r\pi/n)} = \frac{1}{2}\left[1 + i \cot\left(\frac{r\pi}{n}\right)\right],$$

where $r = 0, 1, 2, \ldots, (n - 1)$. **Ans.**

Example 5:

Find all the values of $(1)^{1/4}$.

Solution :

Let $1 = r (\cos \theta + i \sin \theta)$

then $r \cos \theta = 1$ and $r \sin \theta = 0$ whence $r = 1$ and $\theta = 0$.

$\therefore (1)^{1/4} = (\cos 0 + i \sin 0)^{1/4} = [\cos (2n\pi + 0) + i \sin (2n\pi + 0)]^{1/4}$

$= \cos (2n\pi/4) + i \sin (2n\pi/4)$

Giving n the values 0, 1, 2 and 3 the required values are

$\cos 0 + i \sin 0$; $\cos\frac{\pi}{2} + i \sin\frac{\pi}{2}$, $\cos \pi + i \sin \pi$ and $\cos\frac{3\pi}{2} + i \sin\frac{3\pi}{2}$

i.e., $1 + i\,0$; $0 + i\,1$; $-1 + i\,0$ and $0 + i\,(-1)$

i.e., 1; i; -1, and $-i$ *i.e.*, ± 1 and $\pm i$. **Ans.**

Example6:

Find all the values of $(-1)^{1/3}$.

Solution :

$-1 = \cos \pi + i \sin \pi$... $\qquad \because \cos \pi = -1$ and $\sin \pi = 0$

$2\ (-1)^{1/3} = (\cos\pi + i\sin\pi)^{1/3}$

$= [\cos(2n\pi + \pi) + i\sin(2n\pi + \pi)]^{1/3}$

$= \cos\left\{\frac{1}{3}(2n\pi+\pi)\right\} + i\sin\left\{\frac{1}{3}(2n\pi+\pi)\right\}$

Giving n the values 0, 1 and 2, the required values are

$\cos\frac{1}{3}\pi + i\sin\frac{1}{3}\pi,\ \cos\pi + i\sin\pi$ and $\cos\frac{5}{3}\pi + i\sin\frac{5}{3}\pi$

i.e., $\frac{1}{2}+i\frac{1}{2}\sqrt{3},\ -1 + i\,(0)$ and $\frac{1}{2}+i\left(-\frac{1}{2}\sqrt{3}\right)$

i.e., $\frac{1}{2}\left(1+i\sqrt{3}\right),\ -1,$ and $\frac{1}{2}\left(1-i\sqrt{3}\right)$ **Ans.**

Example 7:

Find all the values of $(i)^{1/4}$

Solution :

We can write $i = \cos\frac{1}{2}\pi + i\sin\frac{1}{2}\pi$ ..., $\because \cos\frac{1}{2}\pi = 0$ and $\sin\frac{1}{2}\pi = 1$

$\therefore (i)^{1/4} = \left(\cos\frac{1}{2}\pi + i\sin\frac{1}{2}\pi\right)^{1/4}$

$= \left[\cos\left(2n\pi + \frac{1}{2}\pi\right) + i\sin\left(2n\pi + \frac{1}{2}\pi\right)\right]^{1/4}$

$= \left[\cos\frac{1}{2}(4n\pi+\pi) + i\sin\frac{1}{2}(4n\pi+\pi)\right]^{1/4}$

$= \cos\frac{1}{8}(4n\pi + \pi) + i\sin\frac{1}{8}(4n\pi + \pi)$

Giving n the values 0, 1, 2 and 3 we have the required values as :

$\cos\frac{1}{8}\pi + i\sin\frac{1}{8}\pi;\ \cos\frac{5}{8}\pi + i\sin\frac{5}{8}\pi;$

$\cos\frac{9}{8}\pi + i\sin\frac{9}{8}\pi$ and $\cos\frac{13}{8}\pi + i\sin\frac{13}{8}\pi$

Also $\cos\frac{9}{8}\pi + i\sin\frac{9}{8}\pi = \cos\left(\pi + \frac{1}{8}\pi\right) + i\sin\left(\pi + \frac{1}{8}\pi\right)$

$= -\cos\frac{1}{8}\pi - i\sin\frac{1}{8}\pi$

And similarly $\cos \frac{13}{8}\pi + i \sin \frac{13}{8}\pi = -\cos \frac{5}{8}\pi - i \sin \frac{5}{8}\pi$

∴ The required values are

$\cos \frac{r\pi}{8} + i \sin \frac{r\pi}{8}$ and $-\cos \frac{r\pi}{8} - i \sin \frac{r\pi}{8}$, where r = 1, 5. **Ans.**

Example 8:

Find all the values of $(8i)^{1/3}$.

Solution :

$$\therefore 8i = 8\left[\cos \frac{1}{2}\pi + i \sin \frac{1}{2}\pi\right]$$

$$\therefore \cos \frac{1}{2}\pi = 0 \text{ and } \sin \frac{1}{2}\pi = 1$$

$$\therefore (8i)^{1/3} = 8^{1/3}\left[\cos \frac{1}{2}\pi + i \sin \frac{1}{2}\pi\right]^{1/3}$$

$$= 2\left[\cos\left(2n\pi + \frac{1}{2}\pi\right) + i \sin\left(2n\pi + \frac{1}{2}\pi\right)\right]^{1/3}$$

$$= 2\left[\cos \frac{4n\pi + \pi}{6} + i \sin \frac{4n\pi + \pi}{6}\right]$$

Giving n the values 0, 1 and 2 the required values are

$$2\left[\cos \frac{1}{6}\pi + i \sin \frac{1}{6}\pi\right],\ 2\left[\cos \frac{5}{6}\pi + i \sin \frac{5}{6}\pi\right],$$

and $\quad 2\left[\cos \frac{3}{2}\pi + i \sin \frac{5}{2}\pi\right]$

i.e., $\quad 2\left[\frac{1}{2}\sqrt{3} + i.\frac{1}{2}\right],\ 2\left[-\frac{1}{2}\sqrt{3} + i.\frac{1}{2}\right]$ and $2\{0 + i(-1)\}$

i.e., $\quad 2.\frac{1}{2}\left(\sqrt{3} + i\right),\ 2.\frac{1}{2}\left(-\sqrt{3} + i\right)$ and $-2i$

i.e., $\quad \sqrt{3+i},\ -\sqrt{3+i}$ and $-2i$ **Ans.**

Example 9:

Find all the values of $(32)^{1/5}$.

Solution :

$\because 32 = 32\ (1) = 32\ (\cos 0 + i \sin 0)$

$\therefore (32)^{1/5} = (32)^{1/5}\ (\cos 0 + i \sin 0)^{1/5}$

$= 2\ [\cos (2n\pi + 0) + i \sin (2n\pi + 0)]^{1/5}$

$$= 2\left[\cos\frac{2n\pi}{5} + i \sin\frac{2n\pi}{5}\right].$$

Giving n the values 0, 1, 2, 3 and 4 the required values are

$$2\ [\cos 0 + i \sin 0],\ 2\left[\cos\frac{2\pi}{5} + i \sin\frac{2\pi}{5}\right],\ 2\left[\cos\frac{4\pi}{5} + i \sin\frac{4\pi}{5}\right]$$

$$2\left[\cos\frac{6\pi}{5} + i \sin\frac{6\pi}{5}\right] \text{ and } 2\left[\cos\frac{8\pi}{5} + i \sin\frac{8\pi}{5}\right]$$

$$i.e.,\ 2\ [1 + i.0],\ 2\left[\cos\frac{2\pi}{5} + i \sin\frac{2\pi}{5}\right],\ 2\left[\cos\frac{4\pi}{5} + i \sin\frac{4\pi}{5}\right],$$

$$2\left[\cos\left(2\pi - \frac{4\pi}{5}\right) + i \sin\left(2\pi - \frac{4\pi}{5}\right)\right] \text{ and}$$

$$2\left[\cos\left(2\pi - \frac{2\pi}{5}\right) + i \sin\left(2\pi - \frac{2\pi}{5}\right)\right]$$

$$i.e.,\ 2\ (1),\ 2\left[\cos\frac{2\pi}{5} + i \sin\frac{2\pi}{5}\right],\ 2\left[\cos\frac{4\pi}{5} + i \sin\frac{4\pi}{5}\right],$$

$$2\left[\cos\frac{4\pi}{5} - i \sin\frac{4\pi}{5}\right] \text{ and } 2\left[\cos\frac{2\pi}{5} - i \sin\frac{2\pi}{5}\right]$$

$$i.e.,\ 2,\ 2\left[\cos\frac{2\pi}{5} \pm i \sin\frac{2\pi}{5}\right] \text{ and } \left[\cos\frac{4\pi}{5} \pm i \sin\frac{4\pi}{5}\right].$$ **Ans.**

Example 10:

Find all the values of $(1 + i)^{1/3}$.

Solution :

Let $1 + i = r\ (\cos \theta + i \sin \theta)$

then $r \cos \theta = 1$ and $r \sin \theta = 1$,

whence $r = \sqrt{2}$, $\theta = \pi/4$

$$\therefore 1 + i = \sqrt{2}\left(\cos\frac{\pi}{4} + i\sin\frac{\pi}{4}\right)$$

$$\Rightarrow (2 + i)^{1/3} = \left(\sqrt{2}\right)^{1/3}\left(\cos\frac{\pi}{4} + i\sin\frac{\pi}{4}\right)^{1/3}$$

$$\Rightarrow (1 + i)^{1/3} = 2^{1/6}\left[\cos\left(2n\pi + \frac{\pi}{4}\right) + i\sin\left(2n\pi + \frac{\pi}{4}\right)\right]^{1/3}$$

$$= 2^{1/6}\left[\cos\left(\frac{8n\pi + \pi}{12}\right) + i\sin\left(\frac{8n\pi + \pi}{12}\right)\right]$$

Giving n the values 0, 1 and 2 the required values are

$$2^{1/6}\left[\cos\frac{\pi}{12} + i\sin\frac{\pi}{12}\right],\ 2^{1/6}\left[\cos\frac{3\pi}{4} + i\sin\frac{3\pi}{4}\right]$$

and $\quad 2^{1/6}\left[\cos\frac{17\pi}{12} + i\sin\frac{17\pi}{12}\right]$

i.e., $\quad 2^{1/6}\left[\cos\frac{\pi}{12} + i\sin\frac{\pi}{12}\right],\ 2^{1/6}\left[-\frac{1}{\sqrt{2}} + i\frac{1}{\sqrt{2}}\right]$

and $\quad 2^{1/0}\left[\cos\frac{17\pi}{12} + i\sin\frac{17\pi}{12}\right]$

i.e., $\quad 2^{1/6}\left[\cos\frac{\pi}{12} + i\sin\frac{\pi}{12}\right],\ 2^{-/3}(-1 + i)$

and $\quad 2^{1/6}\left[\cos\frac{17\pi}{12} + i\sin\frac{17\pi}{12}\right]$

i.e., $\quad 2^{-1/3}(-1 + i)$ and $2^{1/6}\left[\cos\frac{r\pi}{12} + i\sin\frac{r\pi}{12}\right]$,

where r = 1, 17. **Ans.**

Example 11:

Prove that n, nth roots of unity form a series in G.P.

Solution :

We can write $1 = \cos 0 + i\sin 0 = \cos 2r\pi + i\sin 2r\pi$

$\therefore (1)^{1/n} = (\cos 2r\pi + i\sin 2r\pi)^{1/n}$

$= \cos \frac{2r\pi}{n} + i \sin \frac{2r\pi}{n}$

Giving r the values 0, 1, 2,..., (n – 1), nth roots of unity are :

$(\cos 0 + i \sin 0), \left(\cos \frac{2\pi}{n} + i \sin \frac{2\pi}{n}\right), \left(\cos \frac{4\pi}{n} + i \sin \frac{4\pi}{n}\right),$

$\ldots, \left[\cos \frac{2(n-1)\pi}{n} + i \sin \frac{2(n-1)\pi}{n}\right]$

i.e., $1, \left[\cos \frac{2\pi}{n} + i \sin \frac{2\pi}{n}\right], \left[\cos \frac{4\pi}{n} + i \sin \frac{4\pi}{n}\right],$

$\left[\cos \frac{6\pi}{n} + i \sin \frac{6\pi}{n}\right], \ldots, \left[\cos \frac{2(n-1)\pi}{n} + i \sin \frac{2(n-1)\pi}{n}\right]$

Let $\cos \frac{2\pi}{n} + i \sin \frac{2\pi}{n} = x$

then $x^2 = \left(\cos \frac{2\pi}{n} + i \sin \frac{2\pi}{n}\right)^2 = \cos \frac{4\pi}{n} + i \sin \frac{4\pi}{n},$

$x^3 = \left(\cos \frac{2\pi}{n} + i \sin \frac{2\pi}{n}\right)^3 = \cos \frac{6\pi}{n} + i \sin \frac{6\pi}{n},$

$x^{n-1} = \left(\cos \frac{2\pi}{n} + i \sin \frac{2\pi}{n}\right)^{n-1}$

$= \left[\cos \frac{2(n-1)\pi}{n} + i \sin \frac{2(n-1)\pi}{n}\right]$

Hence from (i) the n, nth roots of unity are

$1, x, x^2, x^3, \ldots, x^{n-1}$, where $x = \cos \frac{2\pi}{n} + i \sin \frac{2\pi}{n}$.

These roots form G.P. whose common ratio is x

i.e., $\cos \frac{2\pi}{n} + i \sin \frac{2\pi}{n}$. **Hence proved.**

Example 12:

Show that the product of the n, nth roots of unity is $(-1)^{n-1}$.

Solution :

As in above, we can find that the n, nth roots of unity are 1, x, x^2, x^3,..., x^{n-1}, where $x = \cos\frac{2\pi}{n} + i\sin\frac{2\pi}{n}$.

The required product of the roots

$= 1.x.x^2.x^3...x^{n-1}$

$= x^{0+1+2....+(n-1)} = x^{(n-1)/2}$, $\qquad \because \Sigma n = \frac{1}{2}n(n+1)$

$= \left[\cos\frac{2\pi}{n} + i\sin\frac{2\pi}{n}\right]^{(n-1)n/2}$, $\qquad \because x = \cos\frac{2\pi}{n} + i\sin\frac{2\pi}{n}$

$= \cos\left\{\frac{(n-1)\,n}{2}.\frac{2\pi}{n}\right\} + i\sin\left\{\frac{(n-1)\,n}{2}.\frac{2\pi}{n}\right\}$

$= \cos\{(n-1)\pi\} + i\sin\{(n-1)\pi\}$

$= (-1)^{n-1} + i(0)$, $\qquad \because \cos n\pi = (-1)^n, \sin n\pi = 0$

$= (-1)^{n-1}$.

Hence proved.

Example 13:

Solve the equation $x^7 + 1 = 0$.

Solution :

$x^7 + 1 = 0 \Rightarrow x^7 = -1 \Rightarrow x = (-1)^{1/7}$

Hence we are to find out all the values of $(-1)^{1/7}$

Now $-1 = \cos\pi + i\sin\pi$

$\Rightarrow (-1)^{1/7} = (\cos\pi + i\sin\pi)^{1/7}$

$= [\cos(2n\pi + \pi) + i\sin(2n\pi + \pi)]^{1/7}$

$= \cos\frac{1}{7}(2n\pi + \pi) + i\sin\frac{1}{7}(2n\pi + \pi)$

Giving n the values 0, 1, 2, 3, 4, 5 and 6 we get the required values as

$\left(\cos\frac{\pi}{7} + i\sin\frac{\pi}{7}\right), \left(\cos\frac{3\pi}{7} + i\sin\frac{3\pi}{7}\right), \left(\cos\frac{5\pi}{7} + i\sin\frac{5\pi}{7}\right),$

$\left(\cos\frac{7\pi}{7} + i\sin\frac{7\pi}{7}\right), \left(\cos\frac{9\pi}{7} + i\sin\frac{9\pi}{7}\right), \left(\cos\frac{11\pi}{7} + i\sin\frac{11\pi}{7}\right)$

$$\text{and } \left(\cos\frac{13\pi}{7}+i\sin\frac{13\pi}{7}\right) \quad \text{...(i)}$$

Out of these values

$$\cos\frac{7\pi}{7}+i\sin\frac{7\pi}{7} = \cos\pi + i\sin\pi = -1$$

$$\cos\frac{9\pi}{7}+i\sin\frac{9\pi}{7} = \cos\left(2\pi-\frac{5\pi}{7}\right)+i\sin\left(2\pi-\frac{5\pi}{7}\right)$$

$$= \cos\frac{5\pi}{7}-i\sin\frac{5\pi}{7};$$

$$\cos\frac{11\pi}{7}+i\sin\frac{11\pi}{7} = \cos\left(2\pi-\frac{3\pi}{7}\right)+i\sin\left(2\pi-\frac{3\pi}{7}\right)$$

$$= \cos\frac{3\pi}{7}-i\sin\frac{3\pi}{7};$$

$$\cos\frac{13\pi}{7}+i\sin\frac{3\pi}{7} = \cos\left(2\pi-\frac{\pi}{7}\right)+i\sin\left(2\pi-\frac{\pi}{7}\right)$$

$$= \cos\frac{\pi}{7}-i\sin\frac{\pi}{7}.$$

Hence from (i) the required values of $(-1)^{1/7}$ or the roots of the given equation are

$$-1, \left(\cos\frac{\pi}{7}\pm i\sin\frac{\pi}{7}\right), \left(\cos\frac{3\pi}{7}\pm i\sin\frac{3\pi}{7}\right), \left(\cos\frac{5\pi}{7}\pm i\sin\frac{5\pi}{7}\right)$$

i.e., $-1,\ \cos\frac{r\pi}{7}\pm i\sin\frac{r\pi}{7}$, where $r = 1, 3, 5.$ **Ans.**

Example 14:

Solve the equation $x^7 = 2^7$.

Solution :

$$x^7 = 2^7 \Rightarrow x = (2^7)^{1/7} = 2\ (1)^7 \quad \text{...(i)}$$

Now $1 = \cos 0 + i\sin 0 \Rightarrow (1)^{1/7} = [\cos 0 + i\sin 0]^{1/7}$

$$\Rightarrow (1)^{1/7} = [\cos(2n\pi) + i\sin(2n\pi)]^{1/7} = \cos\left(\frac{2}{7}n\pi\right) + i\sin\left(\frac{2}{7}n\pi\right)$$

Giving n the values 0, 1, 2, 3, 4, 5 and 6 we have the seven values of $(1)^{1/7}$ as

$(\cos 0 + i \sin 0)$, $\left(\cos \frac{2}{7}\pi + i \sin \frac{2}{7}\pi\right)$, $\left(\cos \frac{4}{7}\pi + i \sin \frac{4}{7}\pi\right)$,

$\left(\cos \frac{6}{7}\pi + i \sin \frac{6}{7}\pi\right)$, $\left(\cos \frac{8}{7}\pi + i \sin \frac{8}{7}\pi\right)$, $\left(\cos \frac{10}{7}\pi + i \sin \frac{10}{7}\pi\right)$

and $\left(\cos \frac{12}{7}\pi + i \sin \frac{12}{7}\pi\right)$ ...(ii)

Also $\cos \frac{8}{7}\pi + i \sin \frac{8}{7}\pi = \cos\left(2\pi - \frac{6}{7}\pi\right) + i \sin\left(2\pi - \frac{6}{7}\pi\right)$

$= \cos \frac{6}{7}\pi - i \sin \frac{6}{7}\pi$

Similarly $\cos \frac{10}{7}\pi + i \sin \frac{10}{7}\pi = \cos \frac{4}{7}\pi - i \sin \frac{4}{7}\pi$

and $\cos \frac{12}{7}\pi + i \sin \frac{12}{7}\pi = \cos \frac{2}{7}\pi - i \sin \frac{2}{7}\pi$

$\therefore$ From (ii), the values of $(1)^{1/7}$ are

$1, \left(\cos \frac{2}{7}\pi \pm i \sin \frac{2}{7}\pi\right), \left(\cos \frac{4}{7}\pi \pm i \sin \frac{4}{7}\pi\right), \left(\cos \frac{6}{7}\pi \pm i \sin \frac{6}{7}\pi\right)$

Hence from (i), the roots of the given equation are2, 2 $\left(\cos \frac{2}{7}\pi \pm i \sin \frac{2}{7}\pi\right)$; 2 $\left(\cos \frac{4}{7}\pi \pm i \sin \frac{4}{7}\pi\right)$; 2 $\left(\cos \frac{6}{7}\pi \pm i \sin \frac{6}{7}\pi\right)$

$\Rightarrow 2, 2\left(\cos \frac{2r\pi}{7} \pm i \sin \frac{2r\pi}{7}\right)$, where r = 1, 2, 3.

Example 15:

Solve the equation $x^{12} - 1 = 0$ and find which of its roots satisfy the equation $x^4 + x^2 + 1 = 0$.

Solution :

$x^{12} - 1 = 0 \Rightarrow (x^6 - 1)(x^6 + 1) = 0$ ...(i)

Now $x^6 + 1 = 0$ gives $x^6 = -1 \Rightarrow x = (-1)^{1/6}$

$\Rightarrow x = (\cos \pi + i \sin \pi)^{1/6}$ ... $\because -1 = \cos \pi + i \sin \pi$

$= [\cos (2n\pi + \pi) + \sin (2n\pi + \pi)]^{1/6}$

$$= \cos\left(\frac{2n\pi + \pi}{6}\right) + i \sin\left(\frac{2n\pi + \pi}{6}\right)$$

Giving n the values 0, 1, 2, 3, 4 and 5 we have the six roots of the equation $x^6 + 1 = 0$ as :

$$\left(\cos\frac{\pi}{6} + i \sin\frac{\pi}{6}\right), \left(\cos\frac{\pi}{2} + i \sin\frac{\pi}{2}\right), \left(\cos\frac{5\pi}{6} + i \sin\frac{5\pi}{6}\right),$$

$$\left(\cos\frac{7\pi}{6} + i \sin\frac{7\pi}{6}\right), \left(\cos\frac{3\pi}{2} + i \sin\frac{3\pi}{2}\right), \left(\cos\frac{11\pi}{6} + i \sin\frac{11\pi}{6}\right)$$

i.e., $\left(\cos\frac{\pi}{6} + i \sin\frac{\pi}{6}\right)$, i. $\left(\cos\frac{5\pi}{6} + i \sin\frac{5\pi}{6}\right)$, $\left[\cos\left(2\pi - \frac{5\pi}{6}\right)\right.$

$\left.+ i \sin\left(2\pi - \frac{5\pi}{6}\right)\right]$, – i and $\left[\cos\left(2\pi - \frac{\pi}{6}\right) + i \sin\left(2\pi - \frac{\pi}{6}\right)\right]$

i.e., $\left(\cos\frac{\pi}{6} + i \sin\frac{\pi}{6}\right)$, i, $\left(\cos\frac{5\pi}{6} + i \sin\frac{5\pi}{6}\right)$,

$\left(\cos\frac{5\pi}{6} - i \sin\frac{5\pi}{6}\right)$, i and $\left(\cos\frac{\pi}{6} - i \sin\frac{\pi}{6}\right)$

i.e., ± i, $\left(\cos\frac{\pi}{6} \pm i \sin\frac{\pi}{6}\right)$ and $\left(\cos\frac{5\pi}{6} \pm i \sin\frac{5\pi}{6}\right)$ ...(ii)

Again $x^6 - 1 = 0$ gives $x^6 - 1 \Rightarrow x = (1)^{1/6}$

$\Rightarrow x = [(\cos 0 + i \sin 0)]^{1/6}$, $\because = \cos 0 + i \sin 0$

$= [\cos (2n\pi) + i \sin (2n\pi)]^{1/6} = \cos (2n\pi/\pi) + i \sin (2n\pi/6)$

Giving n the values 0, 1, 2, 3, 4 and 5 we have the six roots of the equation $x^6 - 1 = 0$ as

$(\cos 0 + i \sin 0)$, $\left(\cos\frac{\pi}{3} + i \sin\frac{\pi}{3}\right), \left(\cos\frac{2\pi}{3} + i \sin\frac{2\pi}{3}\right)$,

$(\cos \pi + i \sin \pi)$, $\left(\cos\frac{4\pi}{3} + i \sin\frac{4\pi}{3}\right)$ and $\left(\cos\frac{5\pi}{3} + i \sin\frac{5\pi}{3}\right)$

i.e., 1, $\left(\cos\frac{\pi}{3} + i \sin\frac{\pi}{3}\right), \left(\cos\frac{2\pi}{3} + i \sin\frac{2\pi}{3}\right)$, – 1,

$$\left[\cos\left(2\pi-\frac{2\pi}{3}\right)+i\sin\left(2\pi-\frac{2\pi}{3}\right)\right]$$

$$\text{and}\left[\cos\left(2\pi-\frac{\pi}{3}\right)+i\sin\left(2\pi-\frac{\pi}{3}\right)\right]$$

i.e., $1, \left(\cos\frac{\pi}{3}+i\sin\frac{\pi}{3}\right), \left(\cos\frac{2\pi}{3}+i\sin\frac{2\pi}{3}\right), -1,$

$$\left(\cos\frac{2\pi}{3}-i\sin\frac{2\pi}{3}\right) \text{ and } \left(\cos\frac{\pi}{3}-i\sin\frac{\pi}{3}\right)$$

i.e., $\pm 1, \left(\cos\frac{\pi}{3}\pm i\sin\frac{\pi}{3}\right)$ and $\left(\cos\frac{2\pi}{3}\pm i\sin\frac{2\pi}{3}\right)$ ...(iii)

Also $x^6 - 1 = (x^2 - 1)(x^4 + x^2 + 1)$.

$\therefore x^2 - 1 = 0$ gives $x = \pm 1$

$\therefore$ From (iii) the roots of $x^4 + x^2 + 1 = 0$ are

$$\cos\frac{\pi}{3}\pm i\sin\frac{\pi}{3} \text{ and } \cos\frac{2\pi}{3}\pm i\sin\frac{2\pi}{3}$$

i.e., $\cos\frac{\pi}{3}\pm i\sin\frac{\pi}{3}$ and $\cos i\left(\pi-\frac{\pi}{3}\right)\pm i\sin\left(\pi-\frac{\pi}{3}\right)$

i.e., $\cos\frac{\pi}{3}\pm i\sin\frac{\pi}{3}$ and $-\left(\cos\frac{\pi}{3}\pm i\sin\frac{\pi}{3}\right)$

i.e., $\pm\left(\cos\frac{\pi}{3}\pm i\sin\frac{\pi}{3}\right)$

Hence all the roots of the equation $x^{12} - 1 = 0$ from (ii) and (iii) are

$$\pm 1, \pm i, \left(\cos\frac{\pi}{6}\pm i\sin\frac{\pi}{6}\right), \left(\cos\frac{5\pi}{6}\pm i\sin\frac{5\pi}{6}\right),$$

$$\left(\cos\frac{\pi}{3}\pm i\sin\frac{\pi}{3}\right) \text{ and } \left(\cos\frac{2\pi}{3}\pm i\sin\frac{2\pi}{3}\right)$$

i.e., $\pm 1, + i, \left(\cos\frac{\pi}{6}\pm i\sin\frac{\pi}{6}\right), \left[\cos\left(\pi-\frac{\pi}{6}\right)\pm i\sin\left(\pi-\frac{\pi}{6}\right)\right],$

$$\left(\cos\frac{\pi}{3}\pm i\sin\frac{\pi}{3}\right) \text{ and } \left[\cos\left(\pi-\frac{\pi}{3}\right)\pm i\sin\left(\pi-\frac{\pi}{3}\right)\right],$$

$$i.e., \pm 1, \pm i, \left(\cos\frac{\pi}{6} \pm i\sin\frac{\pi}{6}\right), -\left[\cos\frac{\pi}{6} \pm i\sin\frac{\pi}{6}\right],$$

$$\left(\cos\frac{\pi}{3} \pm i\sin\frac{\pi}{3}\right) \text{ and } -\left(\cos\frac{\pi}{3} \pm i\sin\frac{\pi}{3}\right)$$

$$i.e., \pm 1, \pm i, \pm\left(\cos\frac{\pi}{6} \pm i\sin\frac{\pi}{6}\right) \text{ and } \pm\left(\cos\frac{\pi}{3} \pm i\sin\frac{\pi}{3}\right)$$ **Ans.**

Example 16:

Solve the equation $x^7 + x^4 + x^3 + 1 = 0$.

Solution :

$x^7 + x^4 + x^3 + 1 = 0 \Rightarrow x^4 (x^3 + 1) + (x^3 + 1) = 0$

$\Rightarrow \quad (x^4 + 1)(x^3 + 1) = 0$ ...(i)

Then equation $x^4 + 1 = 0$ gives $x^4 = -1 \Rightarrow x = (-1)^{1/4}$

$\Rightarrow x = (\cos\pi + i\sin\pi)^{1/4} = [\cos(2n\pi + \pi + i\sin(2n\pi + \pi)]^{1/4}$

$$= \cos\left(\frac{2n\pi + \pi}{4}\right) + i\sin\left(\frac{2n\pi + \pi}{4}\right)$$

Giving n the values 0, 1, 2 and 3 the roots of the equation $x^4 + 1 = 0$ are :

$$\left(\cos\frac{\pi}{4} + i\sin\frac{\pi}{4}\right), \left(\cos\frac{3\pi}{4} + i\sin\frac{3\pi}{4}\right), \left(\cos\frac{5\pi}{4} + i\sin\frac{5\pi}{4}\right)$$

and $$\left(\cos\frac{7\pi}{4} + i\sin\frac{7\pi}{4}\right)$$

i.e., $$\left(\frac{1}{\sqrt{2}} + i\frac{1}{\sqrt{2}}\right), \left(\frac{-1}{\sqrt{2}} + i\frac{1}{\sqrt{2}}\right), \left[\left(\frac{-1}{\sqrt{2}} + i\frac{-1}{\sqrt{2}}\right)\right]$$

and $$\left[\left(\frac{1}{\sqrt{2}} + i\frac{-1}{\sqrt{2}}\right)\right]$$

i.e., $$\frac{1+i}{\sqrt{2}}, \frac{-1-i}{\sqrt{2}}, \frac{-1+i}{\sqrt{2}} \text{ and } \frac{1-i}{\sqrt{2}} \; i.e., \; \frac{1+i}{\sqrt{2}} \text{ and } \frac{-1+i}{\sqrt{2}}$$ ...(ii)

The equation $x^3 + 1 = 0$, gives $x^3 = -1 \Rightarrow x = (-1)^{1/3}$

$\Rightarrow x = (\cos\pi + i\sin\pi)^{1/3} = [\cos(2n\pi + \pi) + i\sin(2n\pi + \pi)]^{1/3}$

$$= \cos \frac{2n\pi + \pi}{3} + i \sin \frac{2n\pi + \pi}{3}.$$

Giving n the values 0, 1 and 2 we have the roots of the equation $x^3 + 1 = 0$ as

$$\left(\cos \frac{\pi}{3} + i \sin \frac{\pi}{3}\right), (\cos \pi + i \sin \pi) \text{ and } \left(\cos \frac{5\pi}{3} + i \sin \frac{5\pi}{3}\right)$$

i.e., $\left(\frac{1}{2} + i \frac{\sqrt{3}}{2}\right), - 1 \text{ and } \left[\frac{1}{2} + i\left(-\frac{\sqrt{3}}{2}\right)\right]$

i.e., $\frac{1 \pm i\sqrt{3}}{2}$ and $- 1$ ...(iii)

Hence the required roots of given equation from (ii) and (iii) are $\pm \left(\frac{1 \pm i}{\sqrt{2}}\right), - 1$ and $\frac{1 \pm i\sqrt{3}}{2}$. **Ans.**

Example 17:

Solve the equation $x^9 - x^5 + x^4 - 1 = 0$.

Solution :

$$x^9 - x^5 + x^4 - 1 = 0$$

$\Rightarrow x^5 (x^4 - 1) + (x^4 - 1) = 0 \Rightarrow (x^5 + 1)(x^4 - 1) = 0$ ...(i)

The equation $x^4 - 1 = 0$ gives $x^4 = 1 \Rightarrow x^2 = \pm 1$

$\Rightarrow x = \pm \sqrt{1}$ and $\pm \sqrt{-1}$ *i.e.,* ± 1 and $\pm i$. ...(ii)

The equation $x^5 + 1 = 0$ gives $x^5 = - 1 \Rightarrow x = (- 1)^{1/5}$

$$\Rightarrow x = (\cos \pi + i \sin \pi)^{1/5} = [\cos (2n\pi + \pi) + i \sin (2n\pi + \pi)]^{1/5}$$

$$= \cos \left(\frac{2n\pi + \pi}{5}\right) + i \sin \left(\frac{2n\pi + \pi}{5}\right)$$

Giving n the values 0, 1, 2, 3 and 4 we have the roots of the equation $x^5 + 1 = 0$ as :

$$\left(\cos \frac{\pi}{5} + i \sin \frac{\pi}{5}\right), \left(\cos \frac{3\pi}{5} + i \sin \frac{3\pi}{5}\right), (\cos \pi + i \sin \pi),$$

$$\left(\cos \frac{7\pi}{5} + i \sin \frac{7\pi}{5}\right) \text{ and } \left(\cos \frac{9\pi}{5} + i \sin \frac{9\pi}{5}\right)$$

But $\cos\frac{7\pi}{5} + i\sin\frac{7\pi}{5} = \cos\left(2\pi - \frac{3\pi}{5}\right) + i\sin\left(2\pi - \frac{2\pi}{5}\right)$

$= \cos\frac{3\pi}{5} - i\sin\frac{3\pi}{5}$

and $\cos\frac{9\pi}{5} + i\sin\frac{9\pi}{5} = \cos\left(2\pi - \frac{\pi}{5}\right) + i\sin\left(2\pi - \frac{\pi}{5}\right)$

$= \cos\frac{\pi}{5} - i\sin\frac{\pi}{5}$.

Hence the roots of the equation $x^5 + 1 = 0$ are

$\left(\cos\frac{\pi}{5} \pm \sin\frac{\pi}{5}\right), \left(\cos\frac{3\pi}{5} \pm \sin\frac{3\pi}{5}\right)$ and $(\cos\pi + i\sin\pi)$ ...(iii)

Hence from (ii) and (iii) the roots of the given equation are

$\pm 1, \pm i, (\cos\pi + i\sin\pi), \left(\cos\frac{\pi}{5} \pm \sin\frac{\pi}{5}\right)$

and $\left(\cos\frac{3\pi}{5} \pm \sin\frac{3\pi}{5}\right)$ **Ans.**

Example 18:

Solve the equation

$$x^6 + x^5 + x^4 + x^3 + x^2 \quad x + 1 = 0.$$

Solution :

Multiplying the given equation by $(x - 1)$ we have

$(x - 1)(x^6 + x^5 + x^4 + x^3 + x^2 + x + 1) = 0$ **(Note)**

$\Rightarrow x^7 - 1 = 0$

$\Rightarrow x = (1)^{1/7} = (\cos 0 + i\sin 0)^{1/7} = (\cos 2n\pi + i\sin 2n\pi)^{1/7}$

$= \cos\frac{2n\pi}{7} + i\sin\frac{2n\pi}{7}$

Giving n the values 0, 1, 2, 3, 4, 5 and 6 we have the roots of the equation

$(\cos 0 + i\sin 0), \left(\cos\frac{2}{7}\pi + \sin\frac{2}{7}\pi\right), \left(\cos\frac{4}{7}\pi + \sin\frac{4}{7}\pi\right)$

$\left(\cos\frac{6}{7}\pi + \sin\frac{6}{7}\pi\right), \left(\cos\frac{8}{7}\pi + \sin\frac{8}{7}\pi\right), \left(\cos\frac{10}{7}\pi + \sin\frac{10}{7}\pi\right)$

and $\left(\cos\frac{12}{7}\pi + \sin\frac{12}{7}\pi\right)$

But $\cos\frac{8}{7}\pi + i\sin\frac{8}{7}\pi = \cos\frac{6}{7}\pi - i\sin\frac{6}{7}\pi$

$\cos\frac{10}{7}\pi + i\sin\frac{10}{7}\pi = \cos\frac{4}{7}\pi - i\sin\frac{4}{7}\pi$

and $\cos\frac{12}{7}\pi + i\sin\frac{12}{7}\pi = \cos\frac{2}{7}\pi - i\sin\frac{2}{7}\pi$

Hence the roots of the equation $x^7 - 1 = 0$ are

$1, \left(\cos\frac{2}{7}\pi \pm i\sin\frac{2}{7}\pi\right), \left(\cos\frac{4}{7}\pi \pm i\sin\frac{4}{7}\pi\right)$ and $\left(\cos\frac{6}{7}\pi \pm i\sin\frac{6}{7}\pi\right)$

The root 1 corresponds to the factor $x - 1$ and the remaining six roots are those of the given equation.

Hence the roots of the given equation are

$\left(\cos\frac{2r\pi}{7} \pm i\sin\frac{2r\pi}{7}\right)$, where $r = 1, 2, 3$. **Ans.**

Example 19:

Solve $x^5 - x^4 + x^3 - x^2 + x - 1 = 0$.

Solution :

Multiplying the given equation by $(x + 1)$ we get

$$(x + 1)(x^5 - x^4 + x^3 - x^2 + x - 1) = 0 \Rightarrow x^6 - 1 = 0$$

Now do as above.

$1, \cos\frac{2r\pi}{6} \pm i\sin\frac{2r\pi}{6}$, where $r = 1, 2$. **Ans.**

Example 20:

Solve the equation $x^4 - x^3 - x + 1 = 0$.

Solution :

Multiplying the given equation by $(x + 1)$, we get

$$(x + 1)(x^4 - x^3 + x^2 - x + 1) = 0$$

$$\Rightarrow \quad x^5 + 1 = 0$$

$$\Rightarrow \quad x^5 = -1 \Rightarrow x = (-1)^{1/5}$$

$$\Rightarrow x = (\cos \pi + i \sin \pi)^{1/5} = [\cos (2n\pi + \pi) + i \sin (2n\pi + \pi)]^{1/5}$$

$$= \cos\left(\frac{2n\pi + \pi}{5}\right) + i \sin\left(\frac{2n\pi + \pi}{5}\right)$$

Giving n the values 0, 1, 2, 3 and 4 we have the roots of the equation $x^5 + 1 = 0$ as

$$\left(\cos\frac{1}{5}\pi + i \sin\frac{1}{5}\pi\right), \left(\cos\frac{3}{5}\pi + i \sin\frac{3}{5}\pi\right), (\cos \pi + i \sin \pi),$$

$$\left(\cos\frac{7}{5}\pi + i \sin\frac{7}{5}\pi\right) \text{ and } \left(\cos\frac{9}{5}\pi + i \sin\frac{9}{5}\pi\right)$$

But $\cos\frac{9}{5}\pi + i \sin\frac{9}{5}\pi = \cos\left(2\pi - \frac{1}{5}\pi\right) + i \sin\left(2\pi - \frac{1}{5}\pi\right)$

$$= \cos\frac{1}{5}\pi - i \sin\frac{1}{5}\pi,$$

$$\cos\frac{7}{5}\pi + i \sin\frac{7}{5}\pi = \cos\left(2\pi - \frac{3}{5}\pi\right) + i \sin\left(2\pi - \frac{3}{5}\pi\right)$$

$$= \cos\frac{3}{5}\pi - i \sin\frac{3}{5}\pi$$

Hence the roots of the equation $x^5 + 1 = 0$ are

$$\left(\cos\frac{1}{5}\pi \pm i \sin\frac{1}{5}\pi\right), \left(\cos\frac{3}{5}\pi \pm i \sin\frac{3}{5}\pi\right) \text{ and } -1$$

But the root – 1 corresponds to the factor x + 1,

hence the remaining roots are those of the given equation.

Hence the roots of the given equation are

$$\left(\cos\frac{1}{5}\pi \pm i \sin\frac{1}{5}\pi\right) \text{ and } \left(\cos\frac{3}{5}\pi \pm i \sin\frac{3}{5}\pi\right).$$ **Ans.**

Example 21:

Solve that equations $x^{16} - 47x^8 + 1 = 0$.,

Solution :

$$x^{16} - 47x^8 + 1 = 0$$

$$\Rightarrow x^8 = \frac{47 \pm \sqrt{(2209 - 4)}}{2} = \frac{47 \pm \sqrt{(2205)}}{2} = \frac{47 \pm 21\sqrt{5}}{2}$$

$$\Rightarrow x^8 = \left(94 \pm 42\sqrt{5}\right)/4 \text{ (multiplying num. and denom. by 2)}$$

$$= \frac{49 + 45 \pm 2.7\ 3\sqrt{5}}{4} = \frac{\left(7 \pm 3\sqrt{5}\right)^2}{4} = \left[\frac{7 \pm 3\sqrt{5}}{2}\right]^2$$

$$= \left[\frac{14 \pm 6\sqrt{5}}{4}\right]^2, \text{ multiplying num. and denom. by 2}$$

$$= \left[\frac{9 + 5 \pm 2.3\ \sqrt{5}}{4}\right]^2 = \left[\left(\frac{3 \pm \sqrt{5}}{2}\right)^2\right]^2 = \left(\frac{3 \pm \sqrt{5}}{2}\right)^2$$

$$= \left[\frac{6 \pm 2\sqrt{5}}{4}\right]^4, \text{ multiplying num. and denom. by 2}$$

$$= \left[\frac{5 + 1 \pm 2.\sqrt{5.1}}{4}\right]^4 = \left[\left(\frac{\sqrt{5} \pm 1}{2}\right)^2\right]^4 = \left(\frac{\sqrt{5} \pm 1}{2}\right)^8$$

$$\Rightarrow x^8 = \left(\frac{\sqrt{5} \pm 1}{2}\right)^8 (1) = \left(\frac{\sqrt{5} \pm 1}{2}\right)\left[\cos 2n\pi \pm i \sin 2n\pi\right]$$

$$\Rightarrow x = \left(\frac{\sqrt{5} \pm 1}{2}\right)\left[\cos 2n\pi \pm i \sin 2n\pi\right]^{\frac{1}{8}}$$

$$= \left(\frac{\sqrt{5} \pm 1}{2}\right)\left[\cos \frac{n\pi}{4} \pm i \sin \frac{n\pi}{4}\right],$$

where $n = 0, 1, 2$ and 3. **Ans.**

Example 22:

Solve the equation $x^{10} + 11x^5 - 1 = 0$.

Solution :

$$x^{10} + 11x^5 - 1 = 0$$

$$\Rightarrow x^5 = \frac{-11 \pm \sqrt{(121 + 4)}}{2} = \frac{-11 \pm 5\sqrt{5}}{2}$$

$$\Rightarrow x = \frac{-11 \pm 5\sqrt{5}}{2} \times \frac{16}{16} = \frac{-176 \pm 80\sqrt{5}}{32} = \left(\frac{\pm \sqrt{5} - 1}{2}\right)^5$$

$$\Rightarrow x^5 = \left(\frac{\pm \sqrt{5} - 1}{2}\right)^5 \times 1$$ **(Note)**

$$\Rightarrow x^5 = \left(\frac{\pm\sqrt{5}-1}{4}\right)^5 (\cos 2n\pi \pm i \sin 2n\pi)$$

$$\Rightarrow x = \left(\frac{\pm\sqrt{5}-1}{2}\right) [\cos 2n\pi \pm i \sin 2n\pi]^{1/5}$$

$$= \left(\frac{\pm\sqrt{5}-1}{2}\right)\left[\cos \frac{2n\pi}{5} \pm i \sin \frac{2n\pi}{5}\right].$$

Giving n the values 0, 1 and 2 we have the required roots, except for n = 0 we have two equal roots.

Hence the roots of the given equation are given by

$$\frac{\pm\sqrt{5}-1}{2}\left[\cos \frac{2n\pi}{5} \pm i \sin \frac{2n\pi}{5}\right], \text{ where } n = 0, 1, 2.$$ **Ans.**

Example 23:

Solve the equation $(x - 1)^n = x^n$, where n is a positive integer.

Solution :

Since $x \neq 0$ so the given equation can be written as

$$\left(\frac{x-1}{x}\right)^n = 1 = \cos 2r\pi + i \sin 2r\pi$$ **(Note)**

$$\Rightarrow \left(1-\frac{1}{x}\right) = (\cos 2r\pi + i \sin 2r\pi)^{1/n}$$

$$\Rightarrow 1-\frac{1}{x} = \cos\left(\frac{2r\pi}{n}\right) + i \sin\left(\frac{2r\pi}{n}\right), \text{ where } r = 0, 1, 2,..., (n-1)$$

$$\Rightarrow \frac{1}{x} = \left\{1-\cos\left(\frac{2r\pi}{n}\right)\right\} - i\left\{\sin\left(\frac{2r\pi}{n}\right)\right\}$$ **(Note)**

$$= 2 \sin^2 (r\pi/n) - i \{2 \sin (r\pi/n) \cos (r\pi/n)\}$$

$$= 2 \sin (r\pi/n) \{\sin (r\pi/n) - i \cos (r\pi/n)\}$$

$$= \frac{2 \sin(r\pi/n)\{\sin^2(r\pi/n) - i^2 \cos^2(r\pi/n)\}}{\sin(r\pi/n) + i \cos(r\pi/n)},$$ **(Note)**

multiplying num. and denom. by $\sin\left(\frac{r\pi}{n}\right) + i \cos\left(\frac{r\pi}{n}\right)$

$$\Rightarrow \frac{1}{x} = \frac{2\sin(r\pi/n)\{1\}}{\sin(r\pi/n) + i\cos(r\pi/n)}, \because \quad i^2 = -1$$

$$\Rightarrow x = \frac{\sin(r\pi/n) + i\cos(r\pi/n)}{2\sin(r\pi/n)} = \frac{1}{2}\left[1 + i\cot\left(\frac{r\pi}{n}\right)\right],$$

where $r = 0, 1, 2, \ldots, (n-1)$ **Ans.**

Example 24:

Show that roots of the equation

$(1+x)^{2n} + (1-x)^{2n} = 0$ are given by $\pm\, i \tan\dfrac{(2k-1)\pi}{4n}$,

where $k = 1, 2, 3\ldots, n$.

Solution :

The given equation can be written as

$$\left(\frac{1+x}{1-x}\right)^{2n} = -1 = \cos(2r+1)\pi + i\sin(2r+1)\pi \quad \textbf{(Note)}$$

$\Rightarrow y = [\cos(2r+1)\pi + i\sin(2r+1)\pi]^{1/n}$,

where $y\ [(1+x)/(1-x)]^2$ and $r = 0, 1, 2, 3, \ldots, n-1$.

$$\therefore\ y = \cos\left(\frac{2r+1}{n}\right)\pi + i\sin\left(\frac{2r+1}{n}\right)\pi$$

$$\Rightarrow \left(\frac{1+x}{1-x}\right)^2 = \cos\left(\frac{2r+1}{n}\right)\pi + i\sin\left(\frac{2r+1}{n}\right)\pi$$

$$\Rightarrow \frac{1+x}{1-x} = \pm\left[\cos\left(\frac{2r-1}{n}\right)\pi + i\sin\left(\frac{2r+1}{n}\right)\pi\right]^{1/2}$$

$$\Rightarrow \frac{1+x}{1-x} = \pm\cos\left(\frac{2r+1}{2n}\right)\pi \pm i\sin\left(\frac{2r+1}{2n}\right)\pi$$

$$\Rightarrow \frac{(1+x)-(1-x)}{(1+x)+(1-x)} = \frac{\pm\cos\left(\frac{2r+1}{2n}\right)\pi \pm i\sin\left(\frac{2r+1}{2n}\right)\pi - 1}{\pm\cos\left(\frac{2r+1}{2n}\right)\pi \pm i\sin\left(\frac{2r+1}{2n}\right)\pi + 1}$$

$$\Rightarrow x = \frac{\left\{-1 \pm \cos\left(\frac{2r+1}{2n}\right)\pi\right\} \pm i\sin\left(\frac{2r+1}{2n}\right)\pi}{\left\{1 \pm \cos\left(\frac{2r+1}{2n}\right)\pi\right\} \pm i\sin\left(\frac{2r+1}{2n}\right)\pi}$$

$$= \frac{\left\{1 \pm \cos\left(\frac{2r+1}{2n}\right)\pi\right\} \pm i \sin\left(\frac{2r+1}{2n}\right)\pi}{\left\{1 \pm \cos\left(\frac{2r+1}{2n}\right)\pi\right\} \pm i \sin\left(\frac{2r+1}{2n}\right)\pi} \qquad ...(i)$$

Taking upper signs, we get

$$x = \frac{-\left\{2 \sin^2\left(\frac{2r+1}{4n}\right)\pi\right\} + i\, 2 \sin\left(\frac{2r+1}{4n}\right)\pi \cos\left(\frac{2r+1}{4n}\right)\pi}{\left\{2 \cos^2\left(\frac{2r+1}{4n}\right)\pi\right\} + i\, 2 \sin\left(\frac{2r+1}{4n}\right)\pi \cos\left(\frac{2r+1}{4n}\right)\pi}$$

$$= \frac{2i \sin\left(\frac{2r+1}{4n}\right)\pi\left[\cos\left(\frac{2r+1}{4n}\right)\pi + i \sin\left(\frac{2r+1}{4n}\right)\pi\right]}{2 \cos\left(\frac{2r+1}{4n}\right)\pi\left[\cos\left(\frac{2r+1}{4n}\right)\pi + i \sin\left(\frac{2r+1}{4n}\right)\pi\right]}$$

$$= i \tan\left(\frac{2r+1}{4n}\right)\pi = i \tan\left[\frac{2\,(k-1)+1}{4n}\right]\pi, \text{ where } r = k-1$$

$$= i \tan\left(\frac{2k-1}{4n}\right)\pi, \text{ where } k = r + 1 = 1, 2, 3, ..., n \text{ as}$$

$$r = 0, 1, 2, ..., (n-1). \qquad ...(a)$$

Taking lower sign from (ii) we get

$$x = \frac{-2 \cos^2\left(\frac{2r+1}{4n}\right)\pi - i\, 2 \sin\left(\frac{2r+1}{4n}\right)\pi \cos\left(\frac{2r+1}{4n}\right)\pi}{2 \sin^2\left(\frac{2r+1}{4n}\right)\pi - i\, 2 \sin\left(\frac{2r+1}{4n}\right)\pi \cos\left(\frac{2r+1}{4n}\right)\pi}$$

$$= \frac{-2 \cos\left(\frac{2r+1}{4n}\right)\pi\left[\cos\left(\frac{2r+1}{4n}\right)\pi + i \sin\left(\frac{2r+1}{4n}\right)\pi\right]}{-2\, i \sin\left(\frac{2r+1}{4n}\right)\pi\left[\cos\left(\frac{2r+1}{4n}\right)\pi + i \sin\left(\frac{2r+1}{4n}\right)\pi\right]}$$

$$= \frac{1}{i} \cot\left(\frac{2r+1}{4n}\right)\pi = i \cot\left(\frac{2r+1}{4n}\right)\pi$$

$$= = -\tan\left[\frac{\pi}{2} - \frac{(2r+1)\,\pi}{4n}\right], \text{ since } \cot\theta = \tan\left(\frac{\pi}{2} - \theta\right)$$

$$= -i\tan\left[\frac{2n-2r-1}{4n}\right]\pi = -i\tan\left[\frac{2(n-r)-1}{4n}\right]\pi$$

$$= -i\tan\left[\frac{2k-1}{4n}\right]\pi\text{, where } k = n-r \text{ and as } r = 0, 1, 2, (n-1)$$

so $k = n, (n-1), \ldots, 2, 1$

Hence from (a), (b) the required roots are $\pm i\tan\left[\frac{2k-1}{4n}\right]\pi$.

Hence proved.

Example 25:

Find the continued product of the four values of

$$\left(\cos\frac{\pi}{3} + i\sin\frac{\pi}{3}\right)^{3/4}.$$

Solution :

$$\left(\cos\frac{\pi}{3} + i\sin\frac{\pi}{3}\right)^{3/4} = \left[\cos\frac{3\pi}{3} + i\sin\frac{3\pi}{3}\right]^{1/4}$$

$$= [\cos\pi + i\sin\pi]^{1/4}$$

$$= [\cos(2n\pi + \pi) + i\sin(2n\pi + \pi)]^{1/4}$$

$$= \cos\left(\frac{2n\pi + \pi}{4}\right) + i\sin\left(\frac{2n\pi + \pi}{4}\right)$$

Giving n values 0, 1, 2 and 3 we have the four values of

$$\left(\cos\frac{\pi}{3} + i\sin\frac{\pi}{3}\right)^{3/4} \text{ as } \left(\cos\frac{\pi}{4} + i\sin\frac{\pi}{4}\right), \left(\cos\frac{3\pi}{4} + i\sin\frac{3\pi}{4}\right),$$

$$\left(\cos\frac{5\pi}{4} + i\sin\frac{5\pi}{4}\right) \text{ and } \left(\cos\frac{7\pi}{4} + i\sin\frac{7\pi}{4}\right)$$

∴ Required continued product

$$= \left(\cos\frac{\pi}{4} + i\sin\frac{\pi}{4}\right)\left(\cos\frac{3\pi}{4} + i\sin\frac{3\pi}{4}\right)\left(\cos\frac{5\pi}{4} + i\sin\frac{5\pi}{4}\right)$$

$$\times\left(\cos\frac{7\pi}{4} + i\sin\frac{7\pi}{4}\right)$$

$$= \cos\left(\frac{\pi}{4} + \frac{3\pi}{4} + \frac{5\pi}{4} + \frac{7\pi}{4}\right) + i\sin\left(\frac{\pi}{4} + \frac{3\pi}{4} + \frac{5\pi}{4} + \frac{7\pi}{4}\right)$$

$= \cos 4\pi + i \sin 4\pi = 1 + i.0$

$\therefore \cos 4\pi = 1$ and $\sin 4\pi = 0$. **Ans.**

Example 26:

Prove that $(a + ib)^{1/n} + (a - ib)^{1/n}$ has n real values and find those of

$$\left(1+i\sqrt{3}\right)^{1/3} + \left(1-i\sqrt{3}\right)^{1/3}.$$

Solution :

Let $a = r \cos \theta$ and $b = r \sin \theta$,

then $r \sqrt{(a^2+b^2)}$ and $\theta = \tan^{-1} (b/a)$ ...(i)

Substituting these values of a and b we have

$$(a + ib)^{1/n} + (a - ib)^{1/n}$$

$= r^{1/n} (\cos \theta + i \sin \theta)^{1/n} + r^{1/n} (\cos \theta - i \sin \theta)^{1/n}$

$= r^{1/n} [\cos (2r\pi + \theta) + i \sin (2r\pi + \theta)]^{1/n}$

$+ r^{1/n} [\cos (2\pi + \theta) - i \sin (2r\pi + \theta)]^{1/n}$

$$= r^{1/n} \left[\cos\left(\frac{2r\pi +}{n}\right) + \sin\left(\frac{2r\pi+\theta}{n}\right) + \cos\left(\frac{2r\pi+\theta}{n}\right) + i \sin\left(\frac{2rp+\theta}{n}\right)\right]$$

$= 2r^{1/n} \cos \{(2r\pi + \theta)/n\}$

$= 2 (a^2 + b^2)^{1/2n} \cos \dfrac{1}{n}\left(2r\pi + \tan^{-1} \dfrac{b}{a}\right)$, from (i)

which is real and will give n values corresponding to the values 0, 1, 2, 3,...(n – 1) of r.

Putting $a = 1$, $b = \sqrt{3}$ and $n = 3$ we have the three required values of $(1 + i \sqrt{3})^{1/3} + (1 - i\sqrt{3})^{1/3}$ as

$$2 (1 + 3)^{1/6} \cos \frac{1}{3} (2r\pi + \tan^{-1} \sqrt{3}), \text{ where } r = 0, 1, 2$$

i.e., $2 (2^2)^{1/6} \cos \dfrac{1}{3}\left(2r\pi + \dfrac{\pi}{3}\right)$, where $r = 1, 2$

i.e., $2 \times 2^{1/3} \cos \dfrac{1}{3}\left(\dfrac{\pi}{3}\right), 2 \times 2^{1/3} \cos \dfrac{1}{3}\left(\dfrac{7\pi}{3}\right)$

and $\quad 2 \times 2^{1/3} \cos \frac{1}{3}\left(\frac{13\pi}{3}\right)$

i.e., $\quad 2^{4/3} \cos \frac{\pi}{9}, 2^{4/3} \cos \frac{7\pi}{9}$ and $2^{4/3} \cos \frac{13\pi}{9}$

i.e., $\quad 2^{4/3} \cos \frac{1}{9} (r\pi),$

where $\quad r = 1, 7, 13.$ **Ans.**

Example27:

Find all the values of $(-1)^{1/3}$.

Solution :

We have

$-1 = \cos \pi + i \sin \pi$... $\quad \because \cos \pi = -1$ and $\sin \pi = 0$

$2 \ (-1)^{1/3} = (\cos \pi + i \sin \pi)^{1/3}$

$= [\cos(2n\pi + \pi) + i \sin (2n\pi + \pi)]^{1/3}$

$= \cos \left\{\frac{1}{3} (2n\pi + \pi)\right\} + i \sin \left\{\frac{1}{3} (2n\pi + \pi)\right\}$

Giving n the values 0, 1 and 2, the required values are

$\cos \frac{1}{3}\pi + i \sin \frac{1}{3}\pi, \ \cos \pi + i \sin \pi$ and $\cos \frac{5}{3}\pi + i \sin \frac{5}{3}\pi$

i.e., $\quad \frac{1}{2} + i \frac{1}{2} \sqrt{3}, \ -1 + i \ (0)$ and $\frac{1}{2} + i\left(-\frac{1}{2}\sqrt{3}\right)$

i.e., $\quad \frac{1}{2}\left(1 + i\sqrt{3}\right), -1,$ and $\frac{1}{2}\left(1 - i\sqrt{3}\right)$ **Ans.**

$\Rightarrow \quad \cos^2 \alpha + \cos^2 \beta + \cos^2 \gamma = \frac{3}{2}$ **Hence proved.**

Again $\quad \cos 2\alpha + \cos 2\beta + \cos 2\gamma = 0$

$\Rightarrow \quad (1 - 2 \sin^2 \alpha) + (1 - 2 \sin^2 \beta) + (1 - 2 \sin^2 \gamma) = 0$

$\Rightarrow \quad 3 = 2 (\sin^2 \alpha + \sin^2 \beta + \sin^2 \gamma).$

$\Rightarrow \quad \sin^2 \alpha + \sin^2 \beta + \sin^2 \gamma = \frac{3}{2}.$ **Hence proved.**

Example 28:

Express $2 - \sqrt{3 + i}$ in the form $r (\cos \theta + i \sin \theta)$.

Solution :

Let $\left(2-\sqrt{3}\right) + i = r(\cos\theta + i\sin\theta)$ ...(i)

Equating real and imaginary parts on both sides we get

$2-\sqrt{3} = r\cos\theta$ ...(ii), $1 = r\sin\theta$ ...(iii)

Squaring and adding (ii) and (iii) we get

$$r^2 = \left(2-\sqrt{3}\right)^2 + (1)^2 = 4 + 3 - 4\sqrt{3} + 1 = 4\left(2-\sqrt{3}\right)$$

$$= 2\left(4-2\sqrt{3}\right) = 2\left(3+1-2\sqrt{3}\right) = 2\left(\sqrt{3}-1\right)^2$$

$$\therefore r = \sqrt{2}\left(\sqrt{3}-1\right) = \left(\sqrt{6}-\sqrt{2}\right) \qquad \text{...(iv)}$$

Again from (ii) and (iii) we get $\tan\theta = 1/\left(2-\sqrt{3}\right)$

$$\Rightarrow \tan\theta = \frac{2+\sqrt{3}}{\left(2-\sqrt{3}\right)\left(2+\sqrt{3}\right)} = 2+\sqrt{3} \qquad \text{...(v)}$$

∴ With the help of (iv) and (v) we can get the required expression from (i).

Example 29:

Prove that $(\sin x + i\cos x)^n$

$$= \cos n\left(\frac{1}{2}\pi - x\right) + i\sin n\left(\frac{1}{2}\pi - x\right).$$

Solution :

$$(\sin x + i\cos x)^n = \left[\cos\left(\frac{1}{2}\pi - x\right) + i\sin\left(\frac{1}{2}\pi - x\right)\right]^n$$

$$= \cos n\left(\frac{1}{2}\pi - x\right) + i\sin\left(\frac{1}{2}\pi - x\right)$$ **Hence proved.**

3

Expansion of Sin nθ and Cos nθ

3.1 INTRODUCTION

We know by De Moivre's Theorem that

$$(\cos \theta + i \sin \theta)^n = \cos n\theta + i \sin n\theta \qquad \text{...(i)}$$

Also if n be a positive integer, then with the help of Binomial Theorem, we have

$$(\cos \theta + i \sin \theta)^n = c\, x^n\, \theta + {}^nc_1 \cos^{n-1} \theta\, (i \sin \theta) + {}^nc_2 \cos^{n-2} \theta \times (i \sin \theta^2 + {}^nc_3 \cos^{n-3} \theta\, (i \sin \theta)^3 + \ldots$$

$$= \cos^n \theta + i\, {}^nc_1 \cos^{n-1} \theta \sin \theta - {}^nc_2 \cos^{n-2} \theta \sin^2 \theta - i\, {}^nc_3 \cos^{n-3} \theta \sin^3 \theta + \ldots, \text{ since } i = \sqrt{(1)}$$

$$\Rightarrow (\cos \theta + i \sin)^n = [\cos^n \theta - {}^nc_2 \cos^{n-2} \theta \sin^2 \theta + {}^nc_4 \cos^{n-4} \sin^4 \theta - \ldots] + i\, [{}^nc_1 \cos^{n-1} \theta \sin \theta - {}^nc_3 \cos^{n-3} \theta \sin^3 \theta + \ldots] \qquad \text{...(ii)}$$

∴ From (i) and (ii) we have

$$\cos n\theta + i \sin n\theta = [\cos^n \theta - {}^nc_2 \cos^{n-2} \theta \sin^2 \theta + {}^nc_4 \cos^{n-4} \theta \sin^4 \theta - \ldots] + i\, [{}^nc_1 \cos^{n-1} \theta \sin \theta - {}^nc_3 \cos^{n-3} \theta \sin^3 \theta + \ldots]$$

Equating real and imaginary parts on both sides, we have

$$\cos n\theta = \cos^n \cos \theta\ {}^nc_2 \cos^{n-2} \theta \sin^2 \theta + {}^nc_4 \cos^{n-4} \theta \sin^4 \theta - \ldots$$

and $$\sin n\theta = {}^nc_1 \cos^{n-1} \theta \sin \theta - {}^nc_3 \cos^{n-3} \theta \sin^3 \theta + {}^nc_5 \cos^{n-5} \theta \sin^5 \theta - \ldots$$

$$i.e., \cos n\theta = \cos^n\theta - \frac{n(n-1)}{1.2}\cos^{n-2}\theta\sin^2\theta + \frac{n(n-1)(n-2)(n-3)}{1.2.3.4}\cos^{n-4}\theta\sin^4\theta - ...$$

and $$\sin n\theta = n\cos^{n-1}\theta\sin\theta - \frac{n(n-1)(n-2)}{1.2.3}\cos^{n-3}\theta\sin^3\theta + \frac{n(n-1)(n-2)(n-3)(n-4)}{1.2.3.4.5}\cos^{n-5}\theta\sin^5\theta - ...$$

The terms in each of the above series are alternately positive and negative and each series continues till a factor in the numerator vanishes.

Example 1:

Expand cos 4θ.

Solution :

By De Moivre's Theorem,

$$(\cos\theta + i\sin\theta)^4 = \cos 4\theta + i\sin 4\theta \qquad ...(i)$$

Also by Binomial Theorem, we have

$$(\cos\theta + i\sin\theta)^4 = \cos^4\theta + 4\cos^3\theta(i\sin\theta) + \frac{4.3}{1.2}\cos^2\theta(i\sin\theta)^2 + \frac{4.3.2}{1.2.3}\cos\theta(i\sin\theta)^3 + \frac{4.3.2.1}{1.2.3.4}(i\sin\theta)^4$$

$$= \cos\theta + 4i\cos^2\theta\sin\theta - \theta\cos^2\theta\sin^2\theta - 4i\cos\theta\sin^3\theta + \sin^4\theta$$

$$(\cos\theta + \sin\theta)^4 = (\cos^4\theta - 6\cos^2\theta\sin^2\theta + \sin^4\theta) + i(4\cos^3\theta\sin\theta - 4\cos\theta + \sin^3\theta) \qquad ...(ii)$$

∴ From (i) and (ii), we have

$$\cos 4\theta + i\sin 4\theta = (\cos^4\theta - 6\cos^2\theta\sin^2\theta + \sin^4\theta + i(4\cos^3\theta\sin\theta - 4\cos\theta\sin^3\theta)$$

Equating real parts on both sides we get

$$\cos 4\theta = \cos^4\theta - 6\cos^2\theta\sin^2\theta + \sin^4\theta.$$ **Ans.**

Example 2:

Expand cos 7θ.

Solution :

By De Moivre's Theorem,

$$(\cos\theta + i\sin\theta)^7 = (\cos 7\theta + i\sin 7\theta) \qquad ...(i)$$

Also by Bionomial Theorem, we have

$$(\cos\theta + i\sin\theta)^7 = \cos^7\theta + 7c_1\cos^6\theta\,(i\sin\theta) + 7c_2\cos^5\theta\,(i\sin\theta)^2 + 7c_3\cos^4\theta\,(i\sin\theta)^3 + 7c_4\cos^3\theta\,(i\sin\theta)^4 + 7c_5\cos^2\theta\,(i\sin\theta)^5 + 7c_6\cos\theta\,(i\sin\theta)^6 + 7c_7\,(i\sin\theta)^7$$

$$= \cos^7\theta + 7i\cos^6\theta\sin\theta - 21\cos^5\theta\sin^2\theta - 35i\cos^4\theta\sin^3\theta + 35\cos^3\theta\sin^4\theta + 21\,i\cos^2\theta\sin^5\theta - 7\cos\theta\sin^6\theta - i\sin^7\theta$$

$$\Rightarrow (\cos\theta + i\sin\theta)^7 = (\cos^7\theta - 21\cos^5\theta\sin^2\theta + 35\cos^3\theta\sin^4\theta - 7\cos\theta\sin^6\theta) + i\,(7\cos^6\theta\sin\theta - 35\cos^4\theta\sin^3\theta + 21\cos^2\theta\sin^5\theta - \sin^7\theta) \qquad ...(ii)$$

$\therefore$ From (i) and (ii) we get

$$\cos 7\theta + i\sin 7\theta = \cos^7\theta - 21\cos^5\theta\sin^2\theta + 35\cos^3\theta\sin^4\theta - 7\cos\theta\sin^6\theta) + i\,(7\cos^6\sin\theta - 35\cos^4\theta\sin^3\theta + 21\cos^2\theta\sin^5\theta - \sin^7\theta)$$

Equating real parts on both sides we have

$$\cos 7\theta = \cos^7\theta - 21\cos^5\theta\sin^2\theta + 35\cos^3\sin^4\theta - 7\cos\theta\sin^6\theta$$

Example 3:

Expand sin 7θ.

Solution :

By De Moivre's Theorem,

$$(\cos\theta + i\sin\theta)^7 = \cos 7\theta + i\sin 7\theta \qquad ...(i)$$

Also by Binomial Theorem, we have

$$(\cos\theta + i\sin\theta)^7 = \cos^7\theta + 7c_1\cos^6\theta\,(i\sin\theta) + 7c_2\cos^5\theta\,(i\sin\theta)^2 + 7c_3\cos^4\theta\,(i\sin\theta)^3 + 7c_4\cos^3\theta\,(i\sin\theta)^4 + 8c_5\cos^2\theta\,(i\sin\theta)^5 + 7c_6\cos\theta\,(i\sin\theta)^6 + 7c_7\,(i\sin\theta)^7$$

$$= \cos^7\theta + 7i\cos^6\theta\sin\theta - 21\cos^5\theta\sin^2\theta - 35i\cos^4\theta\sin^3\theta + 35\cos^3\theta\sin^4\theta + 21i\cos^2\theta\sin^5\theta\sin^5\theta - 7\cos\theta\sin^6\theta - i\sin^7\theta$$

$$= (\cos^7\theta - 21\cos^5\theta\sin\theta + 35\cos^3\theta\sin^4\theta - 7\cos\theta\sin^6\theta) + i\,(7\cos^6\theta\sin\theta - 35\cos^4\theta\sin^3\theta + 21\cos^2\theta\sin^5\theta - \theta - \sin^7\theta) \qquad ...(ii)$$

∴ From (i) and (ii) we get

$$\cos 7\theta + i \sin 7\theta = (\cos^7\theta - 21 \cos^5\theta \sin^2\theta + + 35 \cos^3\theta \sin^4\theta$$
$$- \cos\theta \sin^6\theta) + i (7 \cos^6\theta \sin\theta - 35 \cos^4\theta \sin^3\theta$$
$$+ 21 \cos^2\theta \sin^5 - \sin^7\theta)$$

Equating imaginary parts on both sides we have

$$\sin 7\theta = 7 \cos^6\theta \sin\theta - 35 \cos^4 \sin^3\theta + 21 \cos^2\theta \sin^5\theta - \sin^7\theta.$$

Ans.

Example 4:

Expand sin 8θ.

Solution :

By De Moivre's Theorem.

$$(\cos\theta + i \sin\theta)^8 = \cos 8\theta + i \sin 8\theta \qquad ...(i)$$

Also by Binomial Theorem, we have

$$(\cos\theta + i \sin\theta)^8 = \cos^8 + 8c_1 \cos^7\theta (i \sin\theta) + 8c_2 \cos^6\theta (i \sin\theta)^2$$
$$+ 8c_3 \cos^5\theta (i \sin\theta)^3 + 8c_4 \cos^4\theta (i \sin\theta)^4 + 8c_5 \cos^3\theta (i \sin\theta)^5$$
$$+ 8c_6 \cos^2\theta (i \sin\theta)^6 + 8c_7 \cos\theta (i \sin\theta)^7 + 8c_8 (i \sin\theta)^8$$
$$= \cos^8\theta + 8i \cos^7\theta \sin\theta - 28 \cos^6\theta \sin^2\theta - 56i \cos^5\theta \sin^3\theta$$
$$+ 70 \cos^4\theta \sin^4\theta + 56i \cos^3\theta \sin^5\theta - 8 \cos^2\theta \sin^6\theta$$
$$- 8i \cos\theta \sin^7\theta + \sin^8\theta$$

$$\Rightarrow (\text{cow}\,\theta + i \sin\theta)^8 = (\cos^8\theta - 28 \cos^6 \sin^2\theta + 70 \cos^4\theta \sin^4\theta$$
$$- 28 \cos^2\theta \sin^6\theta + \sin^8\theta) + i [8 \cos^7\theta \sin\theta - 56 \cos^5\theta \sin^3\theta$$
$$+ 56 \cos^3\theta \sin^5\theta - 8 \cos\theta \sin^7\theta] \quad ...(ii)$$

∴ From (i) and (ii) we have

$$\cos 8\theta + i \sin 8\theta = [\cos^8\theta - 28 \cos^6\theta \sin^2\theta + 70 \cos^4\theta \sin^4\theta$$
$$- 28 \cos^2\theta \sin^6\theta + \sin^8\theta]$$
$$+ i [8 \cos^7\theta \sin\theta - 56 \cos^5\theta \sin^3\theta + 56 \cos^3\theta \sin^5\theta$$
$$- 8 \cos\theta \sin^7\theta]$$

Equating imaginary parts on both sides, we have

$$\sin 8\theta = 8 \cos^7\theta \sin\theta - 56 \cos^5\theta \sin^3\theta + 56 \cos^3\theta \sin^5\theta$$
$$- 8 \cos\theta \sin^7\theta.$$

Ans.

Example 5:

Expand (sin 9θ)/(sin θ) in a series of ascending powers of cos θ.

Solution :

By De Moivre's Theorem, we get

$(\cos \theta + i \sin \theta)^9 = \cos 9\theta + i \sin 9\theta$ ...(i)

Also by Binomial Theorem we have

$(\cos \theta + i \sin \theta)^9 = \cos^9 \theta + 9c_1 \cos^8 \theta (i \sin \theta)$
$+ 9c_2 \cos^7 \theta (i \sin \theta)^2 + 9c_3 \cos^6 \theta (i \sin \theta)^3 + 9c_4 \cos^5 \theta (i \sin \theta)^4$
$+ 9c_5 \cos^4 \theta (i \sin \theta)^5 + 9c_6 \cos^3 \theta (i \sin \theta)^6$
$+ 9c_7 \cos^2 (i \sin \theta)^7 + 9c_8 \cos \theta (i \sin \theta)^8 + 9c_9 (i \sin \theta)^9$
$= \cos^9 + 9i \cos^8 \theta \sin \theta - 36 \cos^7 \theta \sin^2 \theta - 84i \cos^6 \theta \sin^3 \theta$
$+ 126 \cos^5 \theta \sin^4 \theta + 126i \cos^4 \theta \sin^5 \theta - 84 \cos^3 \theta \sin^6 \theta$
$- 36i \cos^2 \theta \sin^7 \theta + 9 \cos \theta \sin^8 \theta + i \sin^9 \theta$
$= (\cos^9 \theta - 36 \cos^7 \theta \sin^2 \theta + 126 \cos^3 \theta \sin^4 \theta - 84 \cos^3 \theta \sin^6 \theta$
$+ 9 \cos \theta \sin^8 \theta) + i (\theta \cos^8 \theta \sin \theta - 84 \cos^6 \theta \sin^3 \theta$
$+ 126 \cos^4 \theta \sin^5 \theta - 36 \cos^2 \theta \sin^7 \theta + \sin^9 \theta)$ ...(ii)

$\therefore$ From (i) and (ii) we have

$\cos 9\theta + i \sin 9\theta = (\cos^9 \theta - 36 \cos^7 \theta \sin^2 \theta + 126 \cos^5 \theta \sin^4 \theta$
$- 84 \cos^3 \theta \sin^6 \theta + 9 \cos \theta \sin^8 \theta) + i (9 \cos^8 \theta \sin \theta$
$- 84 \cos^6 \theta \sin^3 \theta + 126 \cos^4 \theta \sin^5 \theta - 36 \cos^2 \theta \sin^7 \theta + \sin^9 \theta)$

Equating imaginary parts on both sides, we have

$\sin 9\theta = 9 \cos^8 \theta \sin \theta - 84 \cos^6 \theta \sin^3 \theta + 126 \cos^4 \theta \sin^5 \theta$
$- 36 \cos^2 \theta \sin^7 \theta + \sin^9 \theta$

$\Rightarrow \dfrac{\sin 9\theta}{\sin \theta} = 9 \cos^8 \theta - 84 \cos^6 \theta \sin^2 \theta + 126 \cos^4 \theta \sin^4 \theta$
$- 36 \cos^2 \theta \sin^6 \theta + \sin^8 \theta$
$= 9 \cos^8 \theta - 84 \cos^6 \theta (1 - \cos^2 \theta)^3 + 126 \cos^4 \theta (1 - \cos^2 \theta)^2$
$- 36 \cos^2 \theta (1 - \cos^2 \theta)^3 + (1 - \cos^2 \theta)^4, \therefore \sin^2 \theta = 1 - \cos^2 \theta$
$= 9 \cos^8 \theta - 84 \cos^6 \theta + 84 \cos^8 \theta + 126 \cos^4 \theta (1 - 2 \cos^2 \theta$
$+ \cos^4 \theta) - 36 \cos^2 \theta (1 - 3 \cos^2 \theta + 3 \cos^4 \theta - \cos^6 \theta)$
$+ (1 - 4 \cos^2 \theta + 6 \cos^4 \theta - 4 \cos^6 \theta + \cos^8 \theta)$
$= 1 - 40 \cos^2 \theta + 240 \cos^4 \theta - 448 \cos^6 \theta + 256 \cos^8 \theta.$ **Ans.**

Example 6:

When n is odd, prove that the last terms in the expansions of sin nθ and cos nθ are respectively

$(-1)^{(n-1)/2} \sin^n \theta$ *and* $n (-1)^{(n-1)/2} \cos \theta \sin^{n-1} \theta$.

Solution :

If n is odd, then (n – 1) is even.

Now by De Moivre's Theorem

$(\cos\theta + i\sin\theta)^n = \cos n\theta + i\sin n\theta$ ...(i)

Also by Binomial Theorem, we have

$$(\cos\theta + i\sin\theta)^n = \cos^n\theta + {}^nc_1\cos^{n-1}\theta\,(i\sin\theta) + {}^nc_2\cos^{n-2}\theta\,(i\sin\theta)^2 + {}^nc_3\cos^{n-3}\theta\,(i\sin\theta)^3 \ldots$$
$$\ldots + {}^nc_{n-1}\cos\cdot\theta\,(i\sin\theta)^{n-1} + {}^nc_n\,(i\sin\theta)^n\,\theta$$

$$\Rightarrow \cos n\theta + i\sin n\theta = \cos^n\theta + {}^nc_1\, i\cos^{n-1}\theta\sin\theta - {}^nc_2\cos^{n-2}\theta\sin^2\theta - {}^nc_3\, i\cos^{n-3}\theta\sin^3\theta + \ldots + {}^nc_{n-1}i^{n-1}\cos\theta\sin^{n-1}\theta + {}^nc_n\, i^n\sin^n\theta, \text{ from (i)}$$

But $\quad i^{n-1} = (i^2)\,(n-1)^{/2} = (-1)^{(n-1)/2}$

and $\quad i^n = i\times i^{n-1} = i\,(-1)^{(n-1)/2} \qquad \because i^{n-1} = (-1)^{(n-1)/2}$

$$\therefore \cos n\theta + i\sin n\theta = \cos^n\theta + {}^nc_1\, i\cos^{n-1}\theta\sin\theta - {}^nc_2\cos^{n-2}\theta\sin^2\theta - {}^nc_3\, i\cos^{n-3}\theta\sin^3\theta + (-1)^{(n-1)/2}\,{}^nc_{n-1}\cos\theta\sin^{n-1}\theta + i\,(-1)^{(n-1)/2}\,{}^nc_n\sin^n\theta$$

$$= [\cos^n\theta - {}^nc_2\cos^{n-2}\theta\sin^2\theta + \ldots (-1)^{(n-1/2}\,({}^nc_{n-1})\cos\theta\sin^{n-1}\theta] + i\,[{}^nc_1\cos^{n-1}\theta\sin\theta - {}^nc_3\cos^{n-3}\theta\sin^3\theta + \ldots + (-1)^{(n-1)/2}\,{}^nc_n\sin^n\theta]$$

Equating real and imaginary parts both sides, we have

$$\cos n\theta = \cos^n\theta - {}^nc_2\cos^{n-2}\theta\sin^2\theta + \ldots + (-1)^{(n-1)/2}\,{}^nc_{n-1}\cos\theta\sin^n\theta$$

and $$\sin n\theta = {}^nc_1\cos^{n-1}\theta\sin\theta\;{}^nc_3\cos^{n-3}\theta\sin^3\theta + \ldots + (-1)^{(n-1)/2}\,{}^nc_{n-1}\sin^n\theta$$

$\therefore$ The required last terms in the expansions of $\cos n\theta$ and $\sin n\theta$ are $(-1)^{(n-1)/2}\,({}^nc_{n-1})\cos\theta\sin^{n-1}\theta$ and $(-1)^{(n-1)/2}\,{}^nc_n\sin^n\theta$ respectively *i.e.*, $(-1)^{(n-1)/2}.n\cos\theta\sin^{n-1}\theta$ and $(-1)^{(n-1)/2}.1.\sin^n\theta$ respectively

$\because {}^nc_{n-1} = n$ and ${}^nc_n = 1.$

Example 7:

When n is even prove that the last terms in the expansions of sin nθ and cos nθ are respectively

$n\,(-1)^{(n\;2)/2}\cos\theta\sin^{n\;1}\theta$ and $(-1)^{n/2}\sin^n\theta$.

Solution :

As in last example we can prove that

$$\cos n\theta + i \sin n\theta = \cos^n \theta + i.{}^nc_1 \cos^{n-1} \theta \sin \theta - {}^nc_2 \cos^{n-2} \theta \sin^2 \theta - \ldots + {}^nc_{n-1}\, i^{n-1} \cos \theta \sin^{n-1} \theta + {}^nc_n\, i^n \sin^n \theta$$

$$\Rightarrow \cos n\theta + i \sin n\theta = \cos^n \theta + i.n \cos^{n-1} \theta \sin \theta - \frac{n(n-1)}{1.2} \cos^{n-2} \ldots + ni^{n-1} \cos \theta \sin^{n-1} \theta + i^n \sin^n \theta \quad \ldots(i)$$

Now if n is even, n – 2 is also even

Also $i^{n-1} = i \times i^{n-2} = i\,(i^2)^{(n-2)/2}$ **(Note)**

$= i\,(-1)^{(n-2)/2}$, $\because i^2 = -1$

and $i^n = (i^2)^{n/2} = (-1)^{n/2}$, $\because i^2 = -1$

$\therefore$ From (i) substituting these values of i^{n-1} and i^n, we have

$$\cos n\theta + i \sin n\theta = \cos^n \theta + i\, n \cos^{n-1} \theta \sin \theta - \frac{n(n-1)}{1.2} \cos^{n-2} \theta \sin^2 \theta \ldots + ni\,(-1)^{(n-2)/2} \cos^n \theta \sin^{n-1} \theta + (-1)^{n/2} \sin^n \theta$$

$$= \left[\cos^n \theta - \frac{n(n-1)}{1.2} \cos^{n-2} \theta \sin^2 \theta + \ldots + (-1)^{n/2} \sin^n \theta\right]$$

$$+ i\left[n \cos^{n-1} \theta \sin \theta - \frac{n(n-1)(n-2)}{1.2.3} \cos^{n-3} \theta \sin^3 \theta + \ldots + n(-1)^{(n-2)/2} \cos \theta \sin^{n-1} \theta\right]$$

Equating real and imaginary parts on both sides, we get

$$\cos n\theta = \cos^n \theta - \frac{n(n-1)}{1.2} \cos^{n-2} \theta \sin^2 \theta + \ldots + (-1)^{n/2} \sin^n \theta$$

$$\text{and } \sin n\theta = n \cos^{n-1} \theta \sin \theta - \frac{n(n-1)(n-2)}{1.2.3} \cos^{n-3} \theta \sin^3 \theta + \ldots + n(-1)^{(n-2)/2} \cos \theta \sin^{n-1} \theta$$

Hence the required last terms in the expansions of cos nθ and sin nθ are $(-1)^{(n-2)/2}\, n \cos \theta \sin^{n-1} \theta$ respectively.

Example 8:

Prove that

$$\sin n\theta \cos^n \theta = n \tan \theta - \frac{n(n-1)(n-2)}{1.2.3} \tan^3 \theta + \ldots$$

$$\text{and } \cos n\theta \cos^n\theta = 1 - \frac{n(n+1)}{1.2}\tan^2\theta + \frac{n(n+1)(n+2)(n+3)}{1.2.3.4}\tan^4\theta + \ldots$$

Solution :

By De Moivre's Theorem

$$(\cos\theta + i\sin\theta)^{-n} = \cos n\theta - i\sin n\theta \qquad \ldots(i)$$

Also with the help of Binomial Theorem, we have

$(\cos\theta + i\sin\theta)^{-n} = [\cos\theta(1 + i\tan\theta)]^{-n} = (\cos\theta)^{-n}[1 + i\tan\theta]^{-n}$, $\therefore$ the index is negative

$$= \sec^n\theta\left[1 - n(i\tan\theta) + \frac{(-n)(-n-1)}{1.2}(i\tan\theta)^2 + \frac{(-n)(-n-1)(-n-2)}{1.2.3}(i\tan\theta)^3 + \frac{(-n)(-n-1)(-n-2)(-n-3)}{1.2.3.4}(i\tan\theta)^4 + \ldots\right]$$

$$= \sec^n\theta\left[1 - n\,i\tan\theta - \frac{n(n+1)}{1.2}\tan^2\theta + \frac{n(n+1)(n+2)}{1.2.3}i\tan^3\theta + \frac{n(n+1)(n+2)(n+3)}{1.2.3.4}\tan^4\theta - \ldots\right] \qquad \ldots(ii)$$

$\therefore$ From (i) and (ii), we have

$\cos n\theta - i\sin n\theta$

$$= \sec^n\theta\left[\left\{1 - \frac{n(n+1)}{1.2}\tan^2\theta + \frac{n(n+1)(n+2)(n+3)}{1.2.3.4}\tan^4\theta\ldots\right\} - i\left\{n\tan\theta - \frac{n(n+1)(n+2)}{1.2.3}\tan^3\theta + \ldots\right\}\right]$$

Equating real and imaginary parts on both sides we have $\cos n\theta$

$$= \sec^n\theta\left[1 - \frac{n(n+1)}{1.2}\tan^2\theta + \frac{n(n+1)(n+2)(n+3)}{1.2.3.4}\tan^4\theta\ldots\right]$$

$$\text{and} \qquad \sin n\theta = \sec^n\theta\left[n\tan\theta - \frac{n(n+1)(n+2)}{1.2.3}\tan^3\theta + \ldots\right]$$

i.e, $\cos n\theta\cos^n\theta$

$$= 1 - \frac{n(n+1)}{1.2}\tan^2\theta + \frac{n(n+1)(n+2)(n+3)}{1.2.3.4}\tan^4\theta - \ldots$$

and $\sin n\theta \cos^n\theta = n\tan\theta - \dfrac{n(n+1)(n+2)}{1.2.3}\tan^4\theta + \ldots$

Hence proved.

3.2 EXPANSION OF TAN Nθ

In we have proved that

$$\sin n\theta = n\cos^{n-1}\theta\sin\theta - \frac{n(n-1)(n-2)}{1.2.3}\cos^{n-3}\theta\sin^3\theta + \ldots$$

and $\cos n\theta = \cos^n\theta - \dfrac{n(n-1)}{1.2}\cos^{n-2}\sin^2\theta$

$$+ \frac{n(n-1)(n-2)(n-3)}{1.2.3.4}\cos^{n-4}\theta\sin^4\theta - \ldots$$

∴ Dividing we have

$$\tan n\theta = \frac{\sin n\theta}{\cos n\theta}$$

$$= \frac{n\cos^{n-1}\theta\sin\theta - \dfrac{n(n-1)(n-2)}{1.2.3}\cos^{n-3}\sin^3\theta + \ldots}{\cos^n\theta - \dfrac{n(n-1)}{1.2}\cos^{n-2}\sin^2\theta + \dfrac{(n-1)(n-2)(n-3)}{1.2.3.4}\cos^{n-4}\theta\sin^4\theta}$$

Case I. n is odd: The last terms in the expansion of $\sin n\theta$ and $\cos n\theta$ are respectively $(-1)^{(n-1)/2}\sin^n\theta$ and $n(-1)^{(n-1)/2}\cos\theta\sin^{n-1}\theta$.

∴ In the expansion of $\tan n\theta$, the last term in the numerator is $(-1)^{(n-1)/2}\tan^n\theta$ and in the denominator is $n(-1)^{(n-1)/2}\tan^{n-1}\theta$.

Case II. n is even: The last terms in the expansion of $\sin n\theta$ and $\cos n\theta$ are respectively

$n(-1)^{(n-2)/2}\cos\theta\sin^{n-1}\theta$ and $(-1)^{n/2}\sin n\theta$

∴ In the expansion of $\tan n\theta$, the last term in the numerator is $n(-1)^{(n-2/2}\tan^{n-1}\theta$ and in the denominator is $(-1)^{n/2}\tan^n\theta$.

Example 1:

Expand tan 5θ.

Solution :

We have proved above that

$$\tan n\theta = \frac{n\tan\theta - \dfrac{n(n-1)(n-2)}{1.2.3}\tan^3\theta + \ldots}{1 - \dfrac{n(n-1)}{1.2}\tan^2\theta + \dfrac{n(n-1)(n-2)(n-3)}{1.2.3.4}\tan^4\theta - \ldots}$$

$$\therefore \text{ Here } \tan 5\theta = \frac{5\tan\theta - \dfrac{5.4.3}{1.2.3}\tan^3\theta + \dfrac{5.4.3.2.1}{1.2.3.4.5}\tan^5\theta}{1 - \dfrac{5.4}{1.2}\tan^2\theta + \dfrac{5.4.3.2.1}{1.2.3.4.5}\tan^4\theta}$$

$$= \frac{5\tan\theta - 10\tan^3\theta + \tan^5\theta}{1 - 10\tan^2\theta + 5\tan^4\theta}$$ **Ans.**

Example 2:

Expand tan 8θ.

Solution :

We know tan nθ

$$= \frac{n\tan\theta - \dfrac{n(n-1)(n-2)}{1.2.3}\tan^3\theta + \ldots}{1 - \dfrac{n(n-1)}{1.2}\tan^2\theta + \dfrac{n(n-1)(n-2)(n-3)}{1.2.3.4}\tan^4\theta - \ldots}$$

$$\therefore \tan 8\theta = \frac{8\tan\theta - \dfrac{8.7.6}{1.2.3}\tan^2\theta + \dfrac{8.7.6.5.4}{1.2.3.4.5}\tan^5\theta - \dfrac{8.7.6.5.4.3.2}{1.2.3.4.5.6.7}\tan^7\theta}{1 - \dfrac{8.7}{1.2}\tan^2\theta + \dfrac{8.7.6.5}{1.2.3.4}\tan^4\theta - \dfrac{8.7.6.5.4.3}{1.2.3.4.5.6}\tan^6\theta + \dfrac{8.7.6.5.4.3.2.1}{1.2.3.4.5.6.7.8}\tan^8\theta}$$

$$= \frac{8\tan\theta - 55\tan^3\theta + 56\tan^5\theta - 8\tan^7\theta}{1\ 2\tan^2\theta + 70\tan^4\theta - 28\tan^6\theta + \tan^8\theta}$$ **Ans.**

Example 3:

Show that

$$\tan\frac{\theta}{n} + \tan\frac{\theta+\pi}{n} + \tan\frac{\theta+2\pi}{n} + \ldots + \tan\frac{\theta+(n-1)\pi}{n}$$

is equal to n tan θ or – n cot θ according as n is odd or even.

Solution :

Let α denote any one of the angles

$$\frac{\theta}{n}, \frac{\theta+\pi}{n}, \frac{\theta+2\pi}{n}, \ldots, \frac{\theta+(n-1)\pi}{n} \quad \text{...(i)}$$

Then $n\alpha = \theta + m\pi$, where $m = 0, 1, 2, \ldots, (n-1)$

$\Rightarrow \tan n\alpha = \tan(\theta + m\pi) = \tan\theta$ ...(ii)

Case I. n is odd: Then from (ii) we get

$$\frac{{}^nc_1 \tan\alpha - {}^nc_3 \tan^3\alpha + \ldots + (-1)^{(n-1)/2} \tan^n\alpha}{1 - {}^nc_2 \tan^2\alpha + \ldots + (-1)^{(n-1)/2}\, n \tan^{n-1}\alpha} = \tan\theta \quad \textbf{(Note)}$$

$\Rightarrow \tan^n\alpha - n\tan\theta \tan^{n-1}\alpha + \ldots = 0,$ ...(iii)

cross multiplying and simplifying

This equation, given by (iii), is an nth degree equation in $\tan\alpha$ and from (iii) we find that the roots of the equation (ii) are

$$\tan\frac{\theta}{n}, \tan\frac{\theta+\pi}{n}, \tan\frac{\theta+2\pi}{n}, \ldots, \tan\frac{\theta+(n-1)\pi}{n}$$

$\therefore$ Sum of the roots of (iii) is given by

$$\tan\frac{\theta}{n} + \tan\frac{\theta+\pi}{n} + \ldots + \tan\frac{\theta+(n-1)\pi}{n} = n\tan\theta \textbf{ Hence proved.}$$

Case II. n is even: The from (ii), we get

$$\frac{{}^nc_1 \tan\alpha - {}^nc_3 \tan^3\alpha + \ldots + (-1)^{(n-2)/2}\, n \tan^{n-1}\alpha}{1 - {}^nc_2 \tan^2\alpha + \ldots + (-1)^{n/2} \tan^n\alpha}$$

$\Rightarrow \tan^n\alpha \tan\theta + n\tan^{n-1}\alpha + \ldots = 0,$ ...(iv)

cross multiplying and simplifying.

This equation given by (iv) is an nth degree equation in tan a and its roots are the same as in Case I above.

$\therefore$ Sum of the roots of (iv) is given by

$$\tan\frac{\theta}{n} + \tan\frac{\theta+2\pi}{n} + \ldots + \tan\frac{\theta+(n-1)\pi}{n} = -\frac{n}{\tan\theta} = -n\cot\theta$$

Hence proved.

3.3 EXPANSION OF TAN (A + B + Γ + Δ + ...).

Proof : We know

$(\cos\alpha + i\sin\alpha)(\cos\beta + i\sin\beta)(\cos\gamma + i\sin\gamma)(\cos\delta + i\sin\delta)\ldots$

$= \cos(\alpha+\beta+\gamma+\delta+\ldots) + i\sin(\alpha+\beta+\gamma+\delta+\ldots)$...(i)

Now we may write $\cos\alpha + i\sin\alpha = \cos\alpha\,(1 + i\tan\alpha)$

Similarly $\cos\beta + i\sin\beta = \cos\beta\,(1 + i\tan\beta)$,

$\cos\gamma + i\sin\gamma = \cos\gamma\,(1 + i\tan\gamma)$, etc.

Substituting these values in (i) we have

$\cos(\alpha + \beta + \gamma + \delta + \ldots) + i\sin(\alpha + \beta + \gamma + \delta + \ldots)$

$= \cos\alpha\,(1 + i\tan\alpha)\cos\beta\,(1 + i\tan\beta)\ \text{cog}\,\gamma\,(1 + i\tan\gamma)$

$\times \cos\delta\,(1 + i\tan\delta)\ldots$

$= \cos\alpha\cos\beta\cos\gamma\cos\delta\ldots(1 + i\tan\alpha)(1 + i\tan\beta)(1 + i\tan\gamma)$

$\times (1 + i\tan\delta)\ldots$

$= \cos\alpha\cos\beta\cos\gamma\cos\delta\ldots[1 + i(\tan\alpha + \tan\beta + \tan\gamma + \tan\delta + \ldots)$

$+ i^2(\tan\alpha\tan\beta + \tan\alpha\tan\gamma + \tan\beta\tan\gamma + \ldots)$

$+ i^3(\tan\alpha\tan\beta\tan\gamma + \tan\beta\tan\gamma\tan\delta + \ldots) + \ldots]$

$= \cos\alpha\cos\beta\cos\gamma\cos\delta\ldots[1 + is_1 + i^2 s_2 + i^2 s_3 + \ldots]$, ...(ii)

where $s_1 = \Sigma\tan\alpha = \tan\alpha + \tan\beta + \tan\gamma + \tan\delta + \ldots$,

$s_2 = \Sigma\tan\alpha\tan\beta = \tan\alpha\tan\beta + \tan\alpha\tan\gamma + \tan\beta\tan\gamma + \ldots$,

$s_3 = \Sigma\tan\alpha\tan\beta\tan\gamma = \tan\alpha\tan\beta\tan\gamma + \tan\beta\tan\gamma\tan\delta + \ldots$etc.

Hence from (ii) we get

$\cos(\alpha + \beta + \gamma + \delta + \ldots) + i\sin(\alpha + \beta + \gamma + \delta + \ldots)$

$= \cos\alpha\cos\beta\cos\gamma\cos\delta\ldots[1 + is_1 - s_2 - is_3 + s_4 + is_5 - \ldots]$

Equating real and imaginary parts we have

$\cos(\alpha + \beta + \gamma + \delta + \ldots) = \cos\alpha\cos\beta\cos\gamma\cos\delta\ldots[1 - s_2 + s_4 - \ldots]$

and $\sin(\alpha + \beta + \gamma + \delta + \ldots) = \cos\alpha\cos\beta\cos\gamma\cos\delta\ldots$

$[s_1 - s_2 + s_5 + \ldots]$

$\therefore$ Dividing we have,

$$\tan(\alpha+\beta+\gamma+\delta+\ldots) = \frac{\sin(\alpha + \beta + \gamma + \delta + \ldots)}{\cos(\alpha + \beta + \gamma + \delta + \ldots)} = \frac{s_1 - s_2 + s_5 - \ldots}{1 - s_2 + s_4 - \ldots}$$

Example 1:

If α, β, γ be the roots of the equation $x^3 + px^2 + qx + p = 0$ prove that $\tan^{1}\alpha + \tan^{1}\beta + \tan^{1}\gamma = n\pi$ radians except in one particular case.

Solution :

If α, β, γ be the roots of the given equation

$x^3 + px^2 + qx + p = 0$, then we have

$\alpha + \beta + \gamma = -p/1 = -p$

$\alpha\beta + \alpha\gamma + \beta\gamma = \theta/1 = \theta$ and $\alpha\beta\gamma = -p/1 = -p$ ...(i)

Also if $\alpha = \tan\theta_1$, $\beta = \tan\theta_2$ and $\gamma = \tan\theta_3$ then

$$\tan(\theta_1 + \theta_2 + \theta_3) = \frac{"s_1 - s_3"}{1 - s_2}$$

$$= \frac{(\tan\theta_1 + \tan\theta_2 + \tan\theta_3) - (\tan\theta_1 \tan\theta_2 \tan\theta_3)}{1 - (\tan\theta_1 \tan\theta_2 + \tan\theta_1 \tan\theta_2 + \tan\theta_2 \tan\theta_3)}$$

$$= \frac{(\alpha+\beta+\gamma)}{1-(\alpha\beta+\alpha\gamma+\beta\gamma)} = \frac{(-p)-(-p)}{1-q}, \text{ from (i)}$$

$= 0$, except when $\theta = 1$.

$\Rightarrow \theta_1 + \theta_2 + \theta_3 = n\pi$

i.e., $\tan^{-1}\alpha + \tan^{-1}\beta + \tan^{-1}\gamma = n\pi$, except when $q = 1$.

Example 2:

If α, β, γ and δ be the roots of the equation

$x^4 - x^3 \sin 2\theta + x^2 \cos 2\theta\, x \cos\theta \sin\theta = 0$, *prove that*

$$\tan^{-1}\alpha + \tan^{-1}\beta + \tan^{1}\gamma + \tan^{-1}\delta = n\pi + \frac{1}{2}\pi - \theta.$$

Solution :

If α, β, γ and δ be the roots of the equation

$x^4 - x^3 \sin 2\theta + x^2 \cos 2\theta - x\cos\theta - \sin\theta = 0$, then we have

$\alpha + \beta + \gamma + \delta = -(\sin 2\theta)/1 = \sin 2\theta$,

$\alpha\beta + \alpha\gamma + \alpha\delta + \beta\gamma + \beta\delta + \gamma\delta = \cos 2\theta/1 = \cos 2\theta$

$$\alpha\beta\gamma + \alpha\beta\delta + \alpha\gamma\delta + \beta\gamma\delta = \frac{(-\cos\theta)}{1} = \cos\theta$$

and $\alpha\beta\gamma\delta = -\sin\theta$

Let $\alpha = \tan\theta_1$, $\beta = \tan\theta_2$, $\gamma = \tan\theta_3$ and $\delta = \tan\theta_4$, then

$$\tan(\theta_1 + \theta_2 + \theta_3 + \theta_4) = \frac{"s_1 - s_3"}{1 - s_2 + s_4}$$

$$= \frac{(\Sigma \tan\theta_1) - (\Sigma \tan\theta_1 \tan\theta_2 \tan\theta_3)}{1 - (\Sigma \tan\theta_1 \tan\theta_2) + (\Sigma \tan\theta_1 \tan\theta_2 \tan\theta_3\, \theta_4)}$$

$$= \frac{(\Sigma\alpha)\ (\Sigma\alpha\beta\gamma)}{1 - \Sigma\alpha\beta + \Sigma\alpha\beta\gamma} \quad \because \tan\theta_1 = a \text{ etc.}$$

$$= \frac{(\alpha+\beta+\gamma+\delta)-(\alpha\beta\gamma+\alpha\beta\delta+\alpha\gamma\delta+\beta\gamma\delta)}{1-(\alpha\beta+\alpha\gamma+\alpha\delta+\beta\gamma+\beta\delta+\gamma\delta)+\alpha\beta\gamma\delta}$$

$$= \frac{\sin 2\theta - \cos\theta}{1-\cos 2\theta - \sin\theta}, \text{ substituting the values from (i)}$$

$$= \frac{2\sin\theta\cos\theta - \cos\theta}{1-(1-2\sin^2\theta)-\sin\theta} = \frac{\cos\theta(2\sin\theta-1)}{\sin\theta(2\sin\theta-1)} = \cot\theta$$

$$= \tan\left(\frac{1}{2}\pi - \theta\right)$$

$$\Rightarrow \tan(\theta_1+\theta_2+\theta_3+\theta_4) = \tan\left(\frac{1}{2}\pi-\theta\right)$$

$$\Rightarrow \theta_1+\theta_2+\theta_3+\theta_4 = n\pi + \left(\frac{1}{2}\pi-\theta\right)$$

$$\Rightarrow \tan^{-1}\alpha + \tan^{-1}\beta + \tan^{-1}\gamma + \tan^{-1}\delta = n\pi + \left(\frac{1}{2}\pi-\theta\right).$$

Hence proved.

Example 3:

If θ_1, θ_2, θ_3 be three values of θ which satisfy the equation $\tan 2\theta = l \tan(\theta+\alpha)$ and such that no two of them differs by a multiple of π, show that $\theta_1+\theta_2+\theta_3+\alpha$ is a multiple of π.

Solution :

The given equation is $\tan 2\theta = \lambda \tan(\theta+\alpha)$

$$\Rightarrow \quad \frac{2\tan\theta}{1-\tan^2\theta} = \lambda\left(\frac{\tan\theta+\tan\alpha}{1-\tan\theta\tan\alpha}\right)$$

Cross multiplying, we have

$2\tan\theta(1-\tan\theta\tan\alpha) = \lambda(1-\tan^2\theta)(\tan\theta+\tan\alpha)$

$\Rightarrow 2\tan\theta - 2\tan\alpha\tan^2\theta = \lambda(\tan\theta+\tan\alpha-\tan^3\theta-\tan\alpha\tan^2\theta)$

$\Rightarrow \lambda\tan^3\theta - (2-\lambda)\tan^2\alpha\tan\theta + (2-\lambda)\tan\theta - \lambda\tan\alpha = 0,$

which is a cubic equation in $\tan\theta$. Its roots are given as $\tan\theta_1$, $\tan\theta_2$ and $\tan\theta_3$. Therefore we have

$$\tan\theta_1+\tan\theta_2+\tan\theta_3 = -\frac{[(2-\lambda)\tan\alpha]}{\lambda} = \frac{(2-\lambda)\tan\alpha}{\lambda}$$

$\tan\theta_1\tan\theta_2 + \tan\theta_1\tan\theta_3 + \tan\theta_2\tan\theta_3 = (2-l)/l$ and

$\tan\theta_1\tan\theta_2\tan\theta_3 = -[-\lambda\tan\alpha]/i = \tan\alpha$

Also $\tan(\theta_1 + \theta_2 + \theta_3)$

$$= \frac{"s_1 - s_3"}{1 - s_2} = \frac{(\Sigma \tan\theta_1) - (\Sigma \tan\theta_1 \tan\theta_2 \tan\theta_3)}{1 - (\Sigma \tan\theta_1 \tan\theta_2)}$$

$$= \frac{\left[\frac{(2-\lambda)\tan\alpha}{\lambda}\right] - \tan\alpha}{1 - \frac{(2-\lambda)}{\lambda}}$$

$$= \frac{2\tan\alpha - \lambda\tan\alpha - \lambda\tan\alpha}{\lambda - 2 + \lambda} = \frac{(\tan\alpha)(2-2\lambda)}{-(2-2\lambda)} = -\tan\alpha$$

$\Rightarrow \tan(\theta_1 + \theta_2 + \theta_3) = \tan(-\alpha)$

$\Rightarrow \theta_1 + \theta_2 + \theta_3 = n\pi + (-\alpha) = n\pi - \alpha$

$\Rightarrow \theta_1 + \theta_2 + \theta_3 + \alpha = n\pi$. Where n is zero or any integer.

= a multiple of p. **Hence proved.**

Example 4:

Prove that the equation $\cos 2\theta + a\cos\theta + b\sin\theta + c = 0$ has in general, four roots, and that the sum of the four roots is a multiple of 2π.

Solution :

We know $\sin\theta = \dfrac{2\tan\frac{1}{2}\theta}{1 + \tan^2\frac{1}{2}\theta} = \dfrac{2t}{1 + t^2}$, where $t = \tan\frac{1}{2}\theta$

$$\cos\theta = \frac{1 - \tan^2\frac{1}{2}\theta}{1 + \tan^2\frac{1}{2}\theta} = \frac{1 - t^2}{1 + t^2}$$

And $\cos 2\theta = 1 - 2\sin^2\theta = 1 - 2[2t/(1 + t^2)]^2$

$$= \frac{(1+t^2)^2 - 8t^2}{(1+t^2)^2} = \frac{1 - 6t^2 + t^4}{(1+t^2)^2}.$$

Substituting these values in the given equation, we get

$$\frac{1 - 6t^2 + t^4}{(1+t^2)^2} + a\left(\frac{1-t^2}{1+t^2}\right) + b\left(\frac{2t}{1+t^2}\right) + c = 0$$

$\Rightarrow (1 - 6t^2 + t^4) + a(1 - t^2)(1 + t^2) + 2bt(1 + t^2) + c(1 + t^2)^2 = 0$

$\Rightarrow (1 - a + c)t^4 + 2bt^3 + 2(c - 3)t^2 + 2bt + (1 + a + c) = 0.$

The equation is of fourth degree in t or $\tan\frac{1}{2}\theta$, hence it has four roots.

If t_1, t_2, t_3 and t_4 or $\tan\frac{1}{2}\theta_1$, $\tan\frac{1}{2}\theta_2$, $\tan\frac{1}{2}\theta_3$ and $\tan\frac{1}{2}\theta_4$ be its four roots, then

$$\Sigma \tan\frac{1}{2}\theta_1 = \frac{-2b}{1-a+c};\ S \tan\frac{1}{2}\theta_1 \tan\frac{1}{2}\theta_2 = \frac{2(c-3)}{1-a+c};$$

$$\Sigma \tan\frac{1}{2}\theta_1 \tan\frac{1}{2}\theta_2 \tan\frac{1}{2}\theta_3 = \frac{-2b}{1-a+c}$$

and $$\tan\frac{1}{2}\theta_1 \tan\frac{1}{2}\theta_2 \tan\frac{1}{2}\theta_3 \tan\frac{1}{2}\theta_4 = \frac{1+a+c}{1-a+c}$$

Also $$\tan\left(\frac{1}{2}\theta_1+\frac{1}{2}\theta_2+\frac{1}{2}\theta_3+\frac{1}{2}\theta_4\right) = \frac{"s_1-s_3"}{1-s_2+s_4}$$

$$= \frac{\left(\Sigma\tan\frac{1}{2}\theta\right)-\left(\Sigma\tan\frac{1}{2}\theta_1\tan\frac{1}{2}\theta_2\tan\frac{1}{2}\theta_3\right)}{1-\left(\Sigma\tan\frac{1}{2}\theta_1\tan\frac{1}{2}\theta_2\right)+\left(\tan\frac{1}{2}\theta_1\tan\frac{1}{2}\theta_2\tan\frac{1}{2}\theta_3\tan\frac{1}{2}\theta_4\right)}$$

$$= \frac{\left(\frac{-2b}{1-a+c}\right)-\left(\frac{-2b}{1-a+c}\right)}{1-\frac{2(c-5)}{1-a+c}+\frac{1+a+c}{1-a+c}} = 0$$

$$\Rightarrow \frac{1}{2}\theta_1+\frac{1}{2}\theta_2+\frac{1}{2}\theta_3+\frac{1}{2}\theta_4 = n\pi \Rightarrow \theta_1+\theta_2+\theta_3+\theta_4 = 2n\pi$$

$\Rightarrow \theta_1+\theta_2+\theta_3+\theta_4 =$ a multiple of 2π. **Hence proved.**

Example 5:

Prove that the equation $a^2\cos^2\theta + b^2\sin^2\theta + 2ga\cos\theta + 2fb\sin\theta + c = 0$ has four roots and that the sum of the values of θ which satisfy it is an even multiple of π radians.

Or

If θ_1, θ_2, θ_3 and θ_4 be the eccentric angles of the points of intersection of the ellipse $x = a\cos\theta$, $y = b\sin\theta$ and the circle $x^2+y^2+2gx+2fy+c = 0$ show that $\theta_1+\theta_2+\theta_3+\theta_4 = 2n\pi$.

Solution :

We know $$\sin\theta = \frac{2\tan\frac{1}{2}\theta}{1+\tan^2\frac{1}{2}\theta} \text{ and } \cos\theta = \frac{1-\tan^2\frac{1}{2}\theta}{1+\tan^2\frac{1}{2}\theta}$$

Putting $\tan\frac{1}{2}\theta = t$, we have

$\sin\theta = 2t/(1+t^2)$ and $\cos\theta = (1-t^2)/(1+t^2)$ ...(i)

Substituting these values of $\sin\theta$ and $\cos\theta$ in the given equation, it becomes

$$a^2\left(\frac{1-t^2}{1+t^2}\right)^2 + b^2\left(\frac{2t}{1+t^2}\right)^2 + 2ga\left(\frac{1-t^2}{1+t^2}\right) - 2fb\left(\frac{2t}{1+t^2}\right) + c = 0$$

$$\Rightarrow a^2(1-t^2)^2 + b^2(4t^2) + 2ga(1-t^2)(1+t^2) + 2fb(2t)(1+t^2) + c(1+t^2)^2 = 0$$

$$\Rightarrow a^2(1-2t^2+t^4) + b^2(4t^2) + 2ga(1-t^4) + 4fbt(1+t^2) + c(1+2t^2+t^4) = 0$$

$$\Rightarrow t^4(a^2-2ga+c) + 4fbt^3 + t^2(4b^2-2a^2+2c) + 4fbt + (a^2+2ga+c) = 0$$

The equation is of fourth degree in t or $\tan\frac{1}{2}\theta$, hence it has four roots. If t_1, t_2, t_3 and t_4 or $\tan\frac{1}{2}\theta_1$, $\tan\frac{1}{2}\theta_2$, $\tan\frac{1}{2}\theta_3$, $\tan\frac{1}{2}\theta_4$ be its four roots then

$$\Sigma\tan\frac{1}{2}\theta_1 = \frac{-4fb}{a^2+2ga+c^2}\quad \Sigma\tan\frac{1}{2}\theta_1\tan\frac{1}{2}\theta_2 = \frac{4b^2-2a^2+2c}{a^2-2ga+c},$$

$$\Sigma\tan\frac{1}{2}\theta_1\tan\frac{1}{2}\theta_2\tan\frac{1}{2}\theta_3 = \frac{-4fb}{a^2-2ga+c}$$

and $$\tan\frac{1}{2}\theta_1\tan\frac{1}{2}\theta_2\tan\frac{1}{2}\theta_3\tan\frac{1}{2}\theta_4 = \frac{a^2+2ag+c}{a^2-2ga+c}$$

Also $$\left(\tan\frac{1}{2}\theta_1 + \frac{1}{2}\theta_2 + \frac{1}{2}\theta_3 + \frac{1}{2}\theta_4\right) = \frac{\text{"}s_1-s_3\text{"}}{s_2+s_4}$$

$$= \frac{\left(\Sigma\tan\frac{1}{2}\theta_1\right) - \left(\Sigma\tan\frac{1}{2}\theta_1\tan\frac{1}{2}\theta_2\tan\frac{1}{2}\theta_3\right)}{1-\left(\Sigma\tan\frac{1}{2}\theta_1\tan\frac{1}{2}\theta_2\right) + \left(\Sigma\tan\frac{1}{2}\theta_1\tan\frac{1}{2}\theta_2\tan\frac{1}{2}\theta_3\tan\frac{1}{2}\theta_4\right)}$$

$$= \frac{\left(\frac{-4fb}{a^2-2ga+c}\right) - \left(\frac{-4fb}{a^2-2ga+c}\right)}{1-\left(\frac{4b^2-a^2+2c}{a^2-2ga+c}\right) + \left(\frac{a^2+2ga+c}{a^2-2ga+c}\right)} = \frac{0}{\text{some thing}} = 0$$

$$\Rightarrow \tan\left(\frac{1}{2}\theta_1 + \frac{1}{2}\theta_2 + \frac{1}{2}\theta_3 + \frac{1}{2}\theta_4\right) = 0$$

$$\Rightarrow \frac{1}{2}\theta_1 + \frac{1}{2}\theta_2 + \frac{1}{2}\theta_3 + \frac{1}{2}\theta_4 = n\pi$$

$\Rightarrow \theta_1 + \theta_2 + \theta_3 + \theta_4 = n\pi$ = even multiple of π. **Hence proved.**

Example 6:

Prove that the equation as sec θ – bk cosec θ = $a^2 - b^2$, has four roots and that sum of the values of θ, which satisfy it, is equal to an odd multiple of π radians.

Solution :

We know $\sin\theta = \dfrac{2\tan\frac{1}{2}\theta}{1+\tan^2\frac{1}{2}\theta}$ and $\cos\theta = \dfrac{1-\tan^2\frac{1}{2}\theta}{1+\tan^2\frac{1}{2}\theta}$

$\therefore$ cosec $\theta = \dfrac{1+\tan^2\frac{1}{2}\theta}{2\tan\frac{1}{2}\theta}$ and $\sec\theta = \dfrac{1+\tan^2\frac{1}{2}\theta}{1-\tan^2\frac{1}{2}\theta}$

$\Rightarrow$cosec $\theta = \dfrac{1+t^2}{2t}$ and $\sec\theta = \dfrac{1+t^2}{1-t^2}$ where $t = \tan\frac{1}{2}\theta$

Making these substitutions in the given equation we have

$$ah\left(\frac{1+t^2}{1-t^2}\right) - bk\left(\frac{1+t^2}{2t}\right) = a^2 - b^2$$

$\Rightarrow 2ah\,(1 + t^2)\,t - bk\,(1 + t^2)(1 - t^2) = (a^2 - b^2)\,2t\,(1 - t^2)$

$\Rightarrow 2ah\,(t + t^3) - tk\,(1 - t^4) = (a^2 - b^2)(2t - 2t^3)$

$\Rightarrow bk\,t^4 + (2ah + 2a^2 - 2b^2)\,t^3 + (0)\,t^2 + (2ah - 2a^2 + 2b^2)\,t - bk = 0$

This equation is of fourth degree in t_2 hence it has four roots.

Let its roots be t_1, t_2, t_3 and t_4 or $\tan\frac{1}{2}\theta_1$, $\tan\frac{1}{2}\theta_2$, $\tan\frac{1}{2}\theta_3$ and $\tan\frac{1}{2}\theta_4$

Then s_1 = sum of roots taken once

$$= \Sigma\tan\frac{\theta_1}{2} = \frac{-(2ah+2a^2-2b^2)}{bk}$$

s_2 = sum taken two at a time = $\Sigma\tan\dfrac{\theta_1}{2}\tan\dfrac{\theta_2}{2} = \dfrac{0}{bk} = 0$;

s_3 = sum taken three at a time = $\Sigma\tan\frac{1}{2}\theta_1\tan\frac{1}{2}\theta_2\tan\frac{1}{2}\theta_3$

$= -(2ah\ 2a^2 + 2b^2)/bk;$

$$s_4 = \Sigma \tan\frac{1}{2}\theta_1 \tan\frac{1}{2}\theta_2 \tan\frac{1}{2}\theta_3 \tan\frac{1}{2}\theta_4 = -bk/bk = -1$$

$$\text{Also } \tan\left(\frac{1}{2}\theta_1 + \frac{1}{2}\theta_2 + \frac{1}{2}\theta_3 + \frac{1}{2}\theta_4\right) = \frac{"s_1 + s_3"}{1 - s_2 + s_4}$$

$$= \frac{\frac{-(2ah + 2a^2 - 2b^2)}{bk} + \frac{(2ah - 2a^2 + 2h^2)}{bk}}{1 - 0 + (-1)} = \frac{4(b^2 - a^2)}{0}$$

$$= \infty \tan\frac{1}{2}\pi$$

$\Rightarrow \frac{1}{2}\theta_1 + \frac{1}{2}\theta_2 + \frac{1}{2}\theta_3 + \frac{1}{2}\theta_4 = n\pi + \frac{1}{2}\pi$, where n is any integer

$\Rightarrow \theta_1 + \theta_2 + \theta_3 + \theta_4 = 2n\pi + \pi = (2n + 1)\pi$

= odd multiple of p, as $(2n + 1)$ is an odd integer for all integral values of n. **Hence proved.**

4

Expansion of Sin nθ and Cos nθ in Series of Power A

4.1 INTRODUCTION

(i) Expansion of sin α

$$\sin n\theta = n \cos^{n-1}\theta \sin\theta - \frac{n(n-1)(n-2)}{1.2.3}\cos^{n-3}\sin^3\theta + \ldots$$

Putting $n\theta = \alpha \Rightarrow n = \alpha/\theta$ ...(i)

$$\text{we have } \sin\alpha = \frac{\alpha}{\theta}\cos^{n-1}\theta\sin\theta - \frac{\alpha}{\theta}\frac{\left(\frac{\alpha}{\theta}-1\right)\left(\frac{\alpha}{\theta}-2\right)}{1.2.3} \times \cos^{n-3}\sin^3\theta + \ldots$$

$$= \frac{\alpha}{\theta}\cos^{n-1}\theta\sin\theta - \frac{(\alpha-\theta)(\alpha-2\theta)}{\theta^3\, 3!}\cos^{n-3}\theta\sin^3\theta + \ldots$$

$$\Rightarrow \sin\alpha = \alpha\cos^{n-1}\theta\left(\frac{\sin\theta}{\theta}\right) \frac{\alpha(\alpha-\theta)(\alpha-2\theta)}{3!}\cos^{n-3}\theta\left(\frac{\sin\theta}{\theta}\right)^3 + \ldots$$

...(ii)

Let n increase indefinitely, α remaining constant, then from (i) we find θ decreases indefinitely. Consequently (sin θ)/θ and cos θ in the limit when θ is very small are equal to unity and so also their every power is equal to unity. Hence from (ii) we have

$$\sin\alpha = \sin\alpha = \alpha - \frac{\alpha^3}{3!} + \frac{\alpha^5}{5!} - \ldots \text{ ad. inf. (Remember)} \qquad \ldots\text{(iii)}$$

Also if u_n be the nth term of this series (iii), then

$$\lim_{n\to\infty} \frac{u_{n+1}}{|u_n|} = \lim_{n\to\infty} \frac{\alpha^{2n+1}/(2n+1)!}{\alpha^{2n-1}/(2n-1)!} = \lim_{n\to\infty} \frac{\alpha^2}{(2n+1)(2n)}$$
$$= 0 < 1$$

Hence Σu_n or the series (iii) is convergent.

(ii) **Expansion of cos α:** We have proved in that

$$\cos n\theta = \cos^n\theta + \frac{n(n-1)}{1.2}\cos^{n-2}\theta\sin^2\theta + \frac{n(n-1)(n-2)(n-3)}{1.2.3.4} \times \cos^{n-4}\theta\sin^4\theta - \ldots$$

Putting $n\theta = \alpha \Rightarrow n = \alpha/\theta$

$$\cos\alpha = \cos^n\theta - \frac{\frac{\alpha}{\theta}\left(\frac{\alpha}{\theta}-1\right)}{2!}\cos^{n-2}\theta\sin^2\theta + \frac{\frac{\alpha}{\theta}\left(\frac{\alpha}{\theta}-1\right)\left(\frac{\alpha}{\theta}-2\right)\left(\frac{\alpha}{\theta}-3\right)}{4!}\cos^{n-4}\theta\sin^4\theta + \ldots$$

$$= \cos^n\theta - \frac{\alpha(\alpha-\theta)}{\theta^2\, 2!}\cos^{n-2}\theta\sin^2\theta + \frac{\alpha(\alpha-\theta)(\alpha-2\theta)(\alpha-3\theta)}{\theta^4\, 4!} \times \cos^{n-4}\theta\sin^4\theta - \ldots$$

$$\Rightarrow \cos\alpha = \cos^n\theta - \frac{\alpha(\alpha-\theta)}{2!}\cos^{n-2}\theta\left(\frac{\sin\theta}{\theta}\right)^2 + \frac{\alpha(\alpha-\theta)(\alpha-2\theta)(\alpha-3\theta)}{4!}\left(\frac{\sin\theta}{\theta}\right)^4 - \ldots \quad \ldots(i)$$

As in part (i) when n increases indefinitely, α remaining constant, θ decreases indefinitely and consequently (sin θ)/θ and cos θ in the limit when θ is very small are equal to unity and so are every power of them equal to unity.

Hence from (i) above we find that

$$\cos\alpha = 1 - \frac{\alpha^2}{2!} + \frac{\alpha^4}{4!} - \ldots \text{ ad. inf.} \quad \text{(Remember)} \quad \ldots(ii)$$

If u_n be the nth term of this series (ii) then

$$\lim_{n\to\infty} \frac{u_{n+1}}{|u_n|} = \lim_{n\to\infty} \frac{\alpha^{2n}/(2n)\, i}{\alpha^{2n-2}/(2n-2)!} = \lim_{n\to\infty} \frac{\alpha^2}{(2n)(2n-1)}$$
$$= 0 < 1.$$

Hence S u_n or the series (ii) is convergent.

4.2 EXPANSION OF TAN A IN POWERS OF A

$$\tan\alpha = \frac{\sin\alpha}{\cos\alpha} = \frac{\alpha - \frac{\alpha^3}{3!} + \frac{\alpha^5}{5!} - \ldots}{1 - \frac{\alpha^2}{2!} + \frac{\alpha^4}{4!} - \ldots.}$$

$$= \left(\alpha - \frac{\alpha^3}{6} + \frac{\alpha^5}{10} - \ldots\right)\left(1 - \frac{\alpha^2}{2} + \frac{\alpha^4}{24} - \ldots\right)^{-1}$$

$$= \left(\alpha - \frac{\alpha^3}{6} + \frac{\alpha^5}{120} - \ldots\right)\left[1 - \left(\frac{\alpha^2}{2} + \frac{\alpha^4}{24} - \ldots\right)\right]^{-1}$$

$$= \left(\alpha - \frac{\alpha^3}{6} + \frac{\alpha^5}{120} - \ldots\right)\left[1 + \left(\frac{\alpha^2}{2} - \frac{\alpha^4}{24} - \ldots\right) + \left(\frac{a^2}{2} + \frac{a^4}{24} + \ldots\right)^2 + \ldots\right]$$

$$= \left(\alpha - \frac{\alpha^3}{6} + \frac{\alpha^5}{120} - \ldots\right)\left(1 + \frac{\alpha^2}{2} - \frac{\alpha^4}{24} + \frac{\alpha^4}{4} + \ldots\right)$$

$$= \left(\alpha - \frac{\alpha^3}{6} + \frac{\alpha^5}{120} - \ldots\right)\left(1 + \frac{\alpha^2}{2} + \frac{5\alpha^4}{24} + \ldots\right)$$

$$= \alpha + \frac{\alpha^3}{2} + \frac{5\alpha^5}{24} + \ldots - \frac{\alpha^3}{6} - \frac{\alpha^5}{12} - \ldots + \frac{\alpha^5}{120} + \ldots$$

$$= \alpha + \alpha^3\left(\frac{1}{2} - \frac{1}{6}\right) + \alpha^5\left(\frac{1}{24} - \frac{1}{12} + \frac{1}{120}\right) + \ldots$$

$$\Rightarrow \tan\alpha = \alpha + \frac{1}{3}\alpha^3 + \frac{2}{15}\alpha^5 + \ldots \text{ ad. inf.} \qquad \text{(Remember)}$$

Example 1:

If $\frac{\sin\theta}{\theta} = \frac{2165}{2166}$, *prove that* θ *is the number of radians in* 3° *nearly.*

Solution :

$\frac{2165}{2166}$ is nearly equal to. 1.

$\therefore \quad \frac{\sin\theta}{\theta}$ is also nearly equal to 1 and so θ is very small.

$$\text{Now} \quad \frac{\sin\theta}{\theta} = \frac{2165}{2166} \Rightarrow \frac{\theta - (\theta^3/3!) + (\theta^5/5!) - \ldots}{\theta} = \frac{2165}{2166}$$

$$\Rightarrow 1-\frac{\theta^2}{3!}=\frac{2165}{2166}$$, neglecting higher powers of θ as it is very small.

$$\Rightarrow \frac{\theta^2}{3!}=1-\frac{2165}{2166}=\frac{1}{2166} \Rightarrow \quad \theta^2=\frac{6}{2166}=\frac{1}{361}$$

$\Rightarrow \theta = 1/19$ radians

$$=\frac{1}{19}\times\frac{180}{\pi} \text{ degrees,} \qquad \because \pi \text{ radians} = 180^\circ$$

$$=\frac{180\times 7}{19\times 22} \text{ degrees} = 3^\circ \text{ nearly.}$$ **Hence proved.**

Example 2:

If $\frac{\sin\theta}{\theta}=\frac{5765}{5766}$, *find* θ.

Solution :

$\therefore \frac{5765}{5766}$ is nearly equal to 1.

$\therefore \frac{\sin\theta}{\theta}$ is also nearly equal to 1 and so θ is very small.

$$\text{Now } \frac{\sin\theta}{\theta}=\frac{5765}{5766} \Rightarrow \frac{\theta-(\theta^3/3!)+(\theta^5/5!)-\dots}{\theta}=\frac{5765}{5766}$$

$$\Rightarrow 1-\frac{\theta^2}{3!}=\frac{5765}{5766}$$, neglecting higher powers of small θ.

$$\Rightarrow \frac{\theta^2}{6}=1-\frac{5765}{5766}=\frac{1}{5766} \Rightarrow \theta^2=\frac{1}{961}$$

$$\Rightarrow \theta=\frac{1}{31} \text{ radians} = \frac{1}{31}\times\frac{180}{\pi} \text{ degrees,} \therefore \pi \text{ radians} = 180^\circ$$

$$=\frac{180\times 7}{31\times 22} \text{ degrees} = 1.84^\circ \text{ degrees} = 2^\circ \text{ nearly.}$$ **Hence Proved**

Example 3:

If $\cos\theta = \theta$, *prove that* θ *is equal to 0.7 radian nearly.*

Solution :

$$\text{Given } \cos\theta = \theta \Rightarrow 1-\frac{\theta^2}{2!}+\frac{\theta^4}{4!}-\dots = \theta$$

$\Rightarrow 1 - \theta^2/2\,! = 1$, neglecting powers of θ higher than the second.

$\Rightarrow 2 - \theta^2 = 2\theta \Rightarrow \theta^2 + 2\theta - 2 = 0$

$\Rightarrow \theta = \frac{1}{2}\left[-2 + \sqrt{(4+8)}\right] = -1 + \sqrt{3} = 732$ radians,

$= 0.7$ radians nearly. **Hence proved.**

Example 4:

If $\tan\theta = \frac{1}{15}$, *find an approximate value for θ.*

Solution :

If $\tan\theta = \frac{1}{15}$, then $\theta + \frac{\theta^3}{3} + \ldots = \frac{1}{15}$

$\Rightarrow \theta = \frac{1}{15}$, neglecting powers of θ higher than the first

$\Rightarrow \theta = \frac{1}{15}$ radians $= \frac{1}{15} \times \frac{180}{\pi}$ degrees $= \frac{1}{15} \times \frac{180 \times 7}{22}$

$= 42/11 = 3^\circ\, 49'$ nearly. **Ans.**

Example 5:

If $\sin\left(\frac{1}{6}\pi + \theta\right) = 0.51$, *prove that θ is equal to 40' nearly.*

Solution :

$\because$ 0.51 is nearly equal to $\frac{1}{2}$ *i.e.*, $\sin\frac{1}{6}\pi$

$\therefore$ θ must be small.

Now the given equation is $\sin\left(\frac{1}{6}\pi + \theta\right) = 0.51$

$\Rightarrow \sin\frac{1}{6}\pi\cos\theta + \cos\frac{1}{6}\pi\sin\theta = 0.51$

$\Rightarrow \frac{1}{2}\cos\theta + \left(\sqrt{3/2}\right)\sin\theta = 0.51$

$\Rightarrow \frac{1}{2}\left(1 - \frac{\theta^2}{2\,!} + \ldots\right) + \frac{\sqrt{3}}{2}\left(\theta - \frac{\theta^3}{3\,!} + \ldots\right) = 0.51$

$\frac{1}{2}\left(1 - \frac{1}{2}\theta^2\right) + \frac{1}{2}\sqrt{3}\,(\theta) = 0.51$, neglecting powers of θ higher than the second.

$$\Rightarrow -\frac{1}{4}\theta^2 + \frac{1}{2}\sqrt{3q} = .51 - \frac{1}{2} = 0.51 - 0.50 = 0.01 = \frac{1}{100}$$

$$\Rightarrow 25\theta^2 - 50\sqrt{3\theta} + 1 = 0$$

$$\Rightarrow \theta = 0.012 \text{ radian} = \frac{12}{1000} \times \frac{180 \times 7}{22} \times 60 \text{ min} = 40' \text{ nearly.}$$

Example 6:

If $\cos\left(\frac{1}{3}\pi + \theta\right) = 0.49$*, prove that* θ *is equal to 40' nearly.*

Solution :

0.49 is nearly equal to $\frac{1}{2}$ *i.e.,* $\cos \frac{1}{2}\pi$

$\therefore$ θ must be very small.

Now the given equation is $\cos\left(\frac{1}{3}\pi + \theta\right) = 0.49$

$$\Rightarrow \cos\left(\frac{1}{2}\pi\right)\cos\theta + \sin\left(\frac{1}{2}\pi\right)\sin\theta = 0.49$$

$$\Rightarrow \frac{1}{2}\cos\theta - \left(\sqrt{3/2}\right)\sin\theta = 0.49$$

$$\Rightarrow \frac{1}{2}\left(1 - \frac{\theta^2}{2!} + \ldots\right) - \frac{\sqrt{3}}{2}\left(\theta - \frac{\theta^3}{3!} + \ldots\right) = 0.49$$

$\Rightarrow \frac{1}{2}\left(1 - \frac{\theta^2}{2!}\right) - \frac{\sqrt{3}}{2}(\theta) = 0.49$, neglecting powers of θ higher than the second.

$$\frac{1}{2} - \frac{\theta^2}{4} - \frac{\theta\sqrt{3}}{2} = C.49 = \frac{49}{100} \Rightarrow 25\theta^2 + 50\sqrt{3\theta} - 1 = 0$$

$$\Rightarrow \theta = 0.12 \text{ radian nearly} = \frac{12}{1000} \times \frac{180}{22} \times 7 \times 60 \text{ min.}$$

$= 40$ minutes nearly. **Hence proved.**

Example 7:

Calculate the value of sin 3° correct to three significant figures.

Solution :

$$3^\circ = 3 \times \frac{\pi}{180} \text{ radian} = \frac{3 \times 22}{7 \times 180} = \frac{11}{210} \text{ radian}$$

Now we know that $\sin \alpha = \alpha - \frac{\alpha^3}{3!} + \frac{\alpha^5}{5!} + \ldots$

$$\therefore \sin 3^\circ = \sin \frac{11}{210} = \frac{11}{210} - \frac{1}{3\,i}\left(\frac{11}{210}\right)^3 + \ldots$$

$$= .0523 - \frac{1}{6}(.000143) + \ldots$$

$$= .0523 - .000024 + \ldots = .0523.$$ **Ans.**

Example 8:

Prove that

$$\frac{1}{6}\sin^2\theta = \frac{\theta^3}{3!} - \left(1+3^2\right)\frac{\theta^5}{5!} + \left(1+3^2+3^4\right)\frac{\theta^7}{7} - \ldots$$

Solution :

We know that $\sin 3\theta = 3\sin\theta - 4\sin^3\theta$

$$\Rightarrow \quad 4\sin^3\theta = 3\sin\theta - \sin 3\theta$$

$$\Rightarrow \quad \sin^3\theta = \frac{1}{4}[3\sin\theta - \sin 3\theta]$$

$$\Rightarrow \quad \sin^3\theta = \frac{1}{4}\left[3\left\{\theta - \frac{\theta^3}{3!} + \frac{\theta^5}{5!} - \frac{\theta^7}{7!} + \ldots\right\} - \left\{(3\theta) - \frac{(3\theta)^3}{3!} + \frac{(3\theta)^5}{5!} - \frac{(3\theta)^7}{7!} + \ldots\right\}\right]$$

$$= \frac{1}{4}\left[(3\theta - 3\theta) - \frac{\theta^3}{3!}(3-3)^3 + \frac{\theta^5}{5!}(3-3^5) - \frac{\theta^7}{7!}(3-3^7) + \ldots\right]$$

$$= \frac{1}{4}\left[\frac{\theta^3}{3!}(3-3)^3 - \frac{\theta^5}{5!}(3^5-3) + \frac{\theta^7}{7!}(3^7-3) - \ldots\right]$$

$$= \frac{3}{4}\left[\frac{\theta^3}{3!}(3-1) - \frac{\theta^5}{5!}(3^4-1) + \frac{\theta^7}{7!}(3^6-1) - \ldots\right]. \quad \ldots\text{(i)}$$

taking 3 common from each term.

$$= \frac{3}{4}(3^2-1)\left[\frac{\theta^3}{3!} - \frac{\theta^5}{5!}(3^2+1) + \frac{\theta^7}{7!}(3^4+3^2+1) - \ldots\right]$$

$$\sin^3 = \frac{3}{4}\times 8 = \left[\frac{\theta^3}{3!} - \frac{\theta^5}{5!}\left(3^2+1\right) + \frac{\theta^7}{7!}\left(3^4+3^2+1\right) - \ldots\right]$$

$$\Rightarrow \frac{1}{6}\sin^3\theta = \frac{\theta^3}{3!} - \frac{\theta^5}{5!}\left(3^2+1\right) + \frac{\theta^7}{7!}\left(3^4+3^2+1\right) - \ldots$$ **Hence proved.**

Example 9:

Prove that

$$\sin^2\theta\cos\theta = \theta^2 - \frac{5}{6}\theta^4 + \ldots + (-1)^{n+1}\frac{3^{2n}-1}{4}\cdot\frac{\theta^{2n}}{2n!} + \ldots$$

Solution :

$$\sin^2\theta\cos\theta = \sin\theta.\sin\theta\cos\theta = \frac{1}{2}\sin\theta.\sin 2\theta$$

$$= \frac{1}{4}(2\sin\theta\sin 2\theta) = \frac{1}{4}(\cos\theta - \cos 3\theta)$$

$$= \frac{1}{4}\left[\left\{1 - \frac{\theta^2}{2!} + \frac{\theta^4}{4!} - \ldots + (-1)^n\frac{\theta^{2n}}{2n!} + \ldots\right\}\right.$$

$$\left. - \left\{1 - \frac{(3\theta)^2}{2!} + \frac{(3\theta)^4}{4!} - \ldots + \frac{(-1)^n(3\theta)^{2n}}{(2n)!} + \ldots\right\}\right]$$

$$= \frac{1}{4}\left[-\frac{\theta^2}{2!}\left(1-3^2\right) + \frac{\theta^4}{4!}\left(1-3^4\right) - \ldots + (-1)^n\frac{\theta^{2n}}{(2n)!}\left(1-3^{2n}\right) + \ldots\right]$$

$$= \frac{1}{4}\left[4\theta^2 - \frac{10}{3}\theta^4 + \ldots + \frac{(-1)^n(-1)\left(3^{2n}-1\right)\theta^{2n}}{(2n)!} + \ldots\right]$$

$$= \theta^2 - \frac{5}{6}\theta^4 + \ldots + \frac{(-1)^{n+1}\left(3^{2n}-1\right)\theta^{2n}}{(2n)!} + \ldots$$

Example 10:

Expand $(\sin\theta)/\theta$ in powers of θ.

Solution :

$$\frac{\sin\theta}{\theta} = \frac{1}{\theta}\left[\theta - \frac{\theta^2}{3!} + \frac{\theta^5}{5!} - \ldots \text{ ad. anf.}\right]$$

$$= 1 - \frac{\theta^2}{3!} + \frac{\theta^4}{5!} - \frac{\theta^6}{7!} - \ldots \text{ ad. inf.}$$

Ans.

Example 11:

If θ be small, prove that $\theta \cot \theta = 1 - \frac{\theta^2}{3} - \frac{\theta^4}{45}$ *approx.*

Solution :

$$\theta \cot \theta = \frac{\theta}{\tan \theta} = \frac{\theta}{\theta + \frac{1}{3}\theta^3 + \frac{2}{15}\theta^5}, \text{ neglecting higher powers of } \theta.$$

$$\Rightarrow \theta \cot \theta = \frac{\theta}{1 + \frac{1}{3}\theta^2 + \frac{2}{15}\theta^4} = \left[1 + \frac{1}{3}\theta^2 + \frac{2}{15}\theta^4\right]^{-1}$$

$$= 1 - \left(\frac{1}{3}\theta^2 + \frac{2}{15}\theta^4\right) + \left(\frac{1}{2}\theta^2 + \frac{2}{15}\theta^4\right)^2 - ...$$

$$= 1 - \frac{1}{3}\theta^2 - \frac{2}{15}\theta^4 + \frac{1}{9}\theta^4, \text{ neglecting higher powers of } \theta$$

$$= 1 - \frac{1}{3}\theta^2 - \left(\frac{2}{15} - \frac{1}{9}\right)\theta^4 = 1 - \frac{1}{3}\theta^2 - \frac{1}{45}\theta^4 \text{ approx.}$$

Example 12:

Determine a, b, c, if

$$\lim_{\theta \to 0} \frac{\theta\,(a + b\cos\theta) - c\sin\theta}{\theta^5} = 1.$$

Solution :

$$\frac{\theta\,(a + b\cos\theta) - c\sin\theta}{\theta^5} = 1 .$$

$$= \frac{1}{\theta^5}\left[\theta\left\{a + b\left(\theta\frac{\theta^2}{2!} + \frac{\theta^4}{4!} - \frac{\theta^6}{6!} + ...\right)\right\} - c\left\{\theta - \frac{\theta^3}{3!} + \frac{\theta^5}{5!} - \frac{\theta^7}{7!} + ...\right\}\right],$$

expanding cos θ and sin θ

$$= \frac{1}{\theta^5}\left[(a+b+c)\,\theta + \left(\frac{c}{3!} - \frac{b}{2!}\right)\theta^3 + \left(\frac{b}{4!} - \frac{c}{5!}\right)\theta^5 + \left(\frac{c}{7!} - \frac{b}{6!}\right)\theta^7 + ...\right]$$

$$= \frac{(a+b+c)}{\theta^4} + \left(\frac{c}{3!} - \frac{b}{2!}\right)\frac{1}{\theta^2} + \left(\frac{b}{4!} - \frac{c}{5!}\right) + \left(\frac{c}{7!} - \frac{b}{6!}\right)\theta^2 + \ldots$$

$\therefore$ If $\lim\limits_{\theta \to 0} \dfrac{(a + b\cos\theta) - c\sin\theta}{\theta^5} = 1$, then we must have

$$a + b - c = 0, \quad \frac{c}{3!} - \frac{b}{2!} = 0 \text{ and } \frac{b}{4!} - \frac{c}{5!} = 1$$

$\Rightarrow a + b - c = 0;\ c - 3b = 0,\ 5b - c = 120$

Solving these we get a = 120, b = 60, c = 180. **Ans.**

Example 13:

Show that $\dfrac{\pi^2}{2.4} - \dfrac{\pi^4}{2.4.6.8} + \dfrac{\pi^6}{2.4.6.8.10.12} - \ldots = 1$.

Solution :

We know $\cos\theta = 1 - \dfrac{\theta^2}{2!} + \dfrac{\theta^4}{4!} - \dfrac{\theta^6}{6!} + \ldots$

Putting $\theta = \dfrac{1}{2}\pi$ on both sides we have

$$\cos\frac{1}{2}\pi = 1 - \frac{\pi^2}{2^2\, 2!} + \frac{\pi^4}{2^4\, 4!} - \frac{\pi^6}{2^6\, 6!} + \ldots$$

$$\Rightarrow 0 = 1 - \frac{\pi^2}{2 \times 4} + \frac{\pi^4}{16 \times 24} - \frac{\pi^6}{64 \times 720} + \ldots$$

$$\Rightarrow 0 = 1 - \frac{\pi^2}{2.4} + \frac{\pi^4}{2.4.6.8} - \frac{\pi^6}{2.4.6.8.10.12} + \ldots$$

$$\Rightarrow \frac{\pi^2}{2.4} - \frac{\pi^4}{2.4.6.8} + \frac{\pi^6}{2.4.6.8.10.12} - \ldots = 1.$$

Hence proved.

Example 14:

If $x = \dfrac{2}{1!} - \dfrac{4}{3!} + \dfrac{6}{5!} - \dfrac{8}{7!} + \ldots$ *ad. inf.*

and $y = 1 + \dfrac{2}{1!} - \dfrac{2^3}{3!} + \dfrac{2^5}{5!} - \ldots$ *ad. inf., then show that* $x^2 = y$.

Solution :

We know that $\sin\theta = \theta - \dfrac{\theta^3}{3!} + \dfrac{\theta^5}{5!} - \ldots$ ad. inf.

Putting $\theta = 2$ on both sides we have

$$\sin 2 = 2 - \frac{2^3}{3!} + \frac{2^5}{5!} - \dots \text{ ad. inf.}$$

$$\therefore\ y = 1 + \left(\frac{2}{1!} - \frac{2^3}{3!} + \frac{2^5}{5!} - \dots \text{ad. inf.}\right) \Rightarrow y = 1 + \sin 2 \quad \dots(i)$$

Again $x = \frac{2}{1!} - \frac{4}{3!} + \frac{6}{5!} - \frac{8}{7!} + \dots$ ad. inf.

$$= \frac{2+1}{1!} - \frac{3+1}{3!} + \frac{5+1}{5!} - \frac{7+1}{7!} + \dots \text{ ad. inf.}$$

$$= \left(\frac{1}{1!} - \frac{3}{3!} + \frac{5}{5!} - \frac{7}{7!} + \dots \text{ad. anf.}\right) + \left(\frac{1}{1!} - \frac{1}{3!} + \frac{1}{5!} - \frac{1}{7!} + \dots \text{ad. anf.}\right)$$

$$= \left(1 - \frac{1}{2!} + \frac{1}{4!} - \frac{1}{6!} + \dots \text{ad. anf.}\right) + \left(1 - \frac{1}{3!} + \frac{1}{5!} - \frac{1}{7!} + \dots \text{ad. anf.}\right)$$

$= \cos 1 + \sin 1$ [We can verify by putting $\theta = 1$ in the expansions of $\sin \theta$ and $\cos \theta$]

$\therefore\ x^2 = (\cos 1 + \sin 1)^2 = \cos^2 1 + \sin^2 1 + 2 \sin 1 \cos 1$

$= 1 + \sin 2 = y$, from (i)

$\Rightarrow x^2 = y$ **Hence proved.**

Example 15:

Prove that $\left(x - 2\cos\frac{2\pi}{5}\right)\left(x - 2\cos\frac{4\pi}{5}\right)\left(x - 2\cos\frac{6\pi}{5}\right) \times \left(x - 2\cos\frac{8\pi}{5}\right) = x^4 + 2x^3 - x^2 - 2x + 1$

Solution :

$$\left(x - 2\cos\frac{2\pi}{5}\right)\left(x - 2\cos\frac{4\pi}{5}\right)\left(x - 2\cos\frac{6\pi}{5}\right)\left(x - 2\cos\frac{8\pi}{5}\right)$$

$$= \left(x-2\cos\frac{2\pi}{5}\right)\left(x-2\cos\frac{4\pi}{5}\right)\left(x-2\cos\frac{4\pi}{5}\right)\left(x-2\cos\frac{2\pi}{5}\right),$$

$$\because \cos\frac{6\pi}{5} = \cos\left(2\pi - \frac{4\pi}{5}\right) = \cos\frac{4\pi}{5} \text{ and } \cos\frac{8}{5}\pi = \cos\frac{2}{5}\pi$$

$$= \left[\left(x-2\cos\frac{2\pi}{5}\right)\left(x-2\cos\frac{4\pi}{5}\right)\right]^2$$

$$= \left[x^2 - 2x\left(\cos\frac{2\pi}{5} + \cos\frac{4\pi}{5}\right) + 4\cos\frac{2\pi}{5}\cos\frac{4\pi}{5}\right]^2$$

$= [x^2 - 2x (\cos 72^\circ + \cos 144)^\circ + 4 \cos 72^\circ \cos 144^\circ]^2$

$= [x^2 - 2x (\sin 18^\circ - \cos 36^\circ) + 4 (\sin 18^\circ) (- \cos 36^\circ)]^2$

... $\because \cos 72^\circ = \cos (90^\circ - 18^\circ) = \sin 18^\circ$

and $\cos 144^\circ = \cos (\pi - 36^\circ) = - \cos 36^\circ$

$$= \left[x^2 - 2x\left(\frac{\sqrt{5}-1}{4} - \frac{\sqrt{5}+1}{4}\right) + 4\left(\frac{\sqrt{5}-1}{4}\right)\left(-\frac{\sqrt{5}+1}{4}\right)\right]^2$$

$= (x^2 + x - 1)^2 = x^4 + x^2 + 1 + 2x^3 - 2x^3 - 2x^2 - 2x = x^4 + 2x^3 - x^2 - 2x + 1$

Example 16:

(a) Prove that the roots of the equation

$8x^3 - 4x^2 - 4x + 1 = 0$ *are* $\cos\left(\frac{1}{7}\pi\right)$, $\cos\left(\frac{3}{7}\pi\right)$ *and* $\cos\left(\frac{5}{7}\pi\right)$

Hence obtain the equations whose roots are

(b) $\sec^2\left(\frac{1}{7}\pi\right)$, $\sec^2\left(\frac{3}{7}\pi\right)$, $\sec^2\left(\frac{5}{7}\pi\right)$.

(c) $\tan^2\left(\frac{1}{7}\pi\right)$, $\tan^2\left(\frac{3}{7}\pi\right)$, $\tan^2\left(\frac{5}{7}\pi\right)$.

(d) evaluate $\sec\left(\frac{1}{7}\pi\right) + \sec\left(\frac{3}{7}\pi\right) + \sec\left(\frac{5}{7}\pi\right)$.

and evaluate $\sec\left(\frac{1}{7}\pi\right) . \cos\left(\frac{3}{7}\pi\right) . \cos\left(\frac{5}{7}\pi\right)$.

Solution :

(a) Let θ have one of the following seven values

$$\frac{\pi}{7}, \frac{3\pi}{7}, \frac{5\pi}{7}, \frac{7\pi}{7}, \frac{11\pi}{7} \text{ and } \frac{13\pi}{7} \quad \textbf{(Note)} \quad \text{...(i)}$$

i.e., $\theta = \frac{1}{7}(2n+1)\pi$, where $n = 0, 1, 2, ..., 6$

Then $7\theta = (2n+1)\pi \Rightarrow 4\theta = (2n+1)\pi - 3\theta$ **(Note)**

$\Rightarrow \cos 4\theta = \cos[(2n+1)\pi - 3\theta] = \pi - 3\theta] = -\cos 3\theta$ ...(ii) **(Note)**

Now $\cos\theta = 4\cos^3\theta - 3\cos\theta$ and

$$\cos 4\theta = 2\cos^2 2\theta - 1 = 2(2\cos^2\theta - 1)^2 - 1$$

$$= 2(4\cos^4\theta - 4\cos^2\theta + 1) - 1 = 8\cos^4\theta - 8\cos^2\theta + 1$$

Substituting these values of cos 4θ and cos 3θ in (ii) we get

$$8\cos^4\theta - 8\cos^2\theta + 1 = -(4\cos^3\theta - 3\cos\theta)$$

$\Rightarrow \quad 8x^4 + 4x^3 - 8x^2 - 3x + 1 = 0$, where $x = \cos\theta$

$\Rightarrow \quad (x+1)(8x^3 - 4x^2 - 4x + 1) = 0$ ...(iii)

Now x = cos θ and θ can take one of the seven values given in (i).

Also $\quad \cos\frac{13\pi}{7} = \cos\left(2\pi - \frac{\pi}{7}\right) = \cos\frac{\pi}{7};$

$$\cos\frac{11\pi}{7} = \cos\left(2\pi - \frac{3\pi}{7}\right) = \cos\frac{3\pi}{7};$$

and $\quad \cos\frac{9\pi}{7} = \cos\left(2\pi - \frac{5\pi}{7}\right) = \cos\frac{5\pi}{7}.$

Also $\cos\frac{1}{7}(7\pi) = \cos\pi = -1$ *i.e.,* the factor (x + 1) in (iii) corresponds to the value $\frac{1}{7}(7\pi)$ of θ.

Hence the equation whose roots are $\cos\frac{1}{7}\pi$, $\cos\frac{3}{7}\pi$, $\cos\frac{5}{7}\pi$ is

$$8x^3 - 4x^2 - 4x + 1 = 0. \quad \text{...(iv)}$$

(b) In above part we have proved that the roots of the equation $8x^3 - 4x^2 - 4x + 1 = 0$ are $\cos\left(\frac{1}{7}\pi\right)$, $\cos\left(\frac{3}{7}\pi\right)$ and $\cos\left(\frac{5}{7}\pi\right)$

$\therefore \sec^2\frac{\pi}{7} = 1/\cos^2\frac{\pi}{7} \quad \therefore$ Put $y = \frac{1}{x^2}$

$\Rightarrow x = 1/\sqrt{y}$ in the above equation

Then $\sec^2 \frac{1}{7}\pi$, $\sec^2\left(\frac{3}{7}\pi\right)$ and $\sec^2\left(\frac{5}{7}\pi\right)$ are the roots of the equation

$$8\left(1/\sqrt{y}\right)^2 - 4\left(1/\sqrt{y}\right)^2 - 4\left(1/\sqrt{y}\right) + 1 = 0$$

$\Rightarrow \left(8/y\sqrt{y}\right) - (4/y) - \left(4/\sqrt{y}\right)^2 + 1 = 0 \Rightarrow 8 - 4\sqrt{y} - 4y + y\sqrt{y} = 0$

$\Rightarrow 8 - 4y = 4\sqrt{y} - y\sqrt{y} = \sqrt{y}\,(4-y)$

$\Rightarrow (8 - 4y)^2 = y\,(4 - y)^2$ Þ $y^3 - 24y^2 + 80y - 64 = 0.$ **Ans.**

The sum of its roots = 24

i.e., $\sec^2\left(\frac{1}{7}\pi\right) + \sec^2\left(\frac{3}{7}\pi\right) + \sec^2\left(\frac{5}{7}\pi\right) = 24.$

(c) In part (b) above we have proved the equation

$$y^3 - 2y^2 + 80y - 64 = 0$$

has its roots as $\sec^2 \frac{1}{7}\pi$, $\sec^2 \frac{3}{7}\pi$, and $\sec^2 \frac{5}{7}\pi$.

Now we are to obtain the equation whose roots are

$$\tan^2 \frac{1}{7}\pi,\ \tan^2 \frac{3}{7}\pi \text{ and } \tan^2 \frac{5}{7}\pi$$

Now $\tan^2 \frac{1}{7}\pi = \left(\sec^2 \frac{1}{7}\pi\right) - 1$

$\therefore$ Put $z = y - 1 \Rightarrow y = z + 1$ in equation (iii).

The we have required equation as

$$(z + 1)^3 - 24\,(z + 1)^2 + 80\,(z + 1) - 64 = 0$$

$\Rightarrow (z^2 + 3z^2 + 3z + 1) - 24\,(z^2 + 2z + 1) + 80\,(z + 1) - 64 = 0$

$\Rightarrow z^3 - + z^2 + 35z - 7 = 0$ **Ans.**

(d) We have proved in part (a) result (ii) that the roots of the² equation

$8x^3 - 4x^2 - 4x + 1 = 0$ are $\cos \frac{1}{7}\pi$, $\cos \frac{3}{7}\pi$, $\cos \frac{5}{7}\pi$

$$\therefore \cos \frac{1}{7}\pi + \cos \frac{3}{7}\pi + \cos \frac{5}{7}\pi = \frac{4}{8} = \frac{1}{2}$$

$$\cos \frac{1}{7}\pi \cos \frac{3}{7}\pi + \cos \frac{1}{7}\pi \cos \frac{3}{7}\pi \cos \frac{5}{7}\pi = -\frac{4}{8} = -\frac{1}{2}$$

and $\cos\frac{1}{7}\pi\cos\frac{5}{7}\pi\cos\frac{3}{7}\pi = -\frac{1}{8}$

$$\therefore \sec\frac{1}{7}\pi + \sec\frac{3}{7}\pi + \sec\frac{5}{7}\pi = \frac{1}{\cos\frac{1}{7}\pi} + \frac{1}{\cos\frac{3}{7}\pi} + \frac{1}{\cos\frac{5}{7}\pi}$$

$$= \frac{\cos\frac{3}{7}\pi\cos\frac{5}{7}\pi + \cos\frac{1}{7}\pi\cos\frac{5}{7}\pi + \cos\frac{1}{7}\pi\cos\frac{3}{7}\pi}{\cos\frac{1}{7}\pi\cos\frac{3}{7}\pi\cos\frac{5}{7}\pi}$$

$$= \frac{-1/2}{-1/8} = \frac{8}{2} = 4$$ **Ans.**

(e) $\cos\left(\frac{1}{7}\pi\right).\cos\left(\frac{3}{7}\pi\right).\cos\left(\frac{5}{7}\pi\right)$ = product of the roots of the equation (iv).

$$= \frac{1}{8}$$ **Ans.**

Example 17:

Prove that $\sin\frac{\pi}{14}$ *is a root of the equation*

$64z^6 - 80z^4 + 24z^2 - 1 = 0.$

Solution :

We have proved in that the equation whose roots are $\cos\frac{1}{7}\pi$, $\cos\frac{3}{7}\pi$ and $\cos\frac{5}{7}\pi$ is

$$8x^3 - 4x^2 - 4x + 1 = 0 \qquad \text{...(i)}$$

Now as $\cos\frac{1}{7}\pi = 1 - 2\sin^2(\pi/14)$

$\therefore$ Put $x = 1 - 2z^2$ in equation (i) to obtain the required equation.

Hence the required equation is

$$8(1 - 2z^2)^3 - 4(1 - 2z^2)^2 - 4(1 - 2z^2) + 1 = 0$$

$$\Rightarrow 8(1 - 6z^2 + 12z^4 - 8z^6) - 4(1 - 4z^2 + 4z^4) - 4(1 - 2z^2) + 1 = 0$$

$$\Rightarrow 8 - 8z^2 + 96z^4 - 64z^6 - 4 + 16z^2 - 16z^4 - 4 + 8z^2 + 1 = 0$$

$$\Rightarrow 64z^6 - 80z^4 + 24z^2 - 1 = 0.$$ **Hence proved.**

Example 18:

Prove that $\cos\frac{2}{7}\pi$, $\cos\frac{4}{7}\pi$, $\cos\frac{8}{7}\pi$ *are the roots of the equation* $8x^3 + 4x^2 - 4x - 1 = 0$.

Solution :

Let θ denote one of the seven angles

$$\frac{2}{7}\pi, \frac{4}{7}\pi, \frac{6}{7}\pi, \frac{8}{7}\pi, \frac{10}{7}\pi, \frac{12}{7}\pi, \frac{14}{7}\pi \quad \textbf{(Note)} \qquad ..(i)$$

i.e., $\theta = \frac{2}{7}n\pi$, where n = 1, 2, 3,...., 7

$\Rightarrow 7\theta = 2n\pi \quad \Rightarrow \quad 4\theta = 2n\pi - 3\theta$

$\Rightarrow \cos 4\theta = \cos(2n\pi - 3\theta) = \cos 3\theta$ **(Note)**

$\Rightarrow 8\cos^4\theta - 8\cos^2\theta + 1 = 4\cos^3\theta - 3\cos\theta$

$\Rightarrow 8x^4 - 4x^3 - 8x^2 + 3x + 1 = 0$, where $x = \cos\theta$

$\Rightarrow (x - 1)(8x^3 + 4x^2 - 4x - 1) = 0$ ...(ii)

The roots of this equation are x = cow θ, where θ can take any one of the values given in (i).

Also $\cos\frac{14}{7}\pi = \cos 2\pi = 1$ and $(x - 1)$ is the factor corresponding to the value $\frac{14}{7}\pi$ of θ. Hence the roots of the equation $8x^3 + 4x^3 - 4x - 1 = 0$ are the remaining six values of θ in (i). But we know that

$$\cos\frac{12}{7}\pi = \cos\left(2\pi - \frac{2}{7}\pi\right) = \cos\frac{2}{7}\pi; \cos\frac{10}{7}\pi = \cos\left(2\pi - \frac{4}{7}\pi\right) = \cos\frac{4}{7}\pi$$

and $\cos\frac{6}{7}\pi = \cos\left(2\pi - \frac{8}{7}\pi\right) = \cos\frac{8}{7}\pi$. ...(a)

Hence the roots of $8x^3 + 4x^2 - 4x - 1 = 0$ are $\cos\frac{2}{7}\pi$, $\cos\frac{4}{7}\pi$ and $\cos\frac{8}{7}\pi$.

Hence proved.

Example 19:

In the above example find the equations whose rots are $\sec^2\frac{2}{7}\pi$, $\sec^2\frac{4}{7}\pi$, $\sec^2\frac{6}{7}\pi$.

Solution :

From above we find that the equation whose roots are

$\cos\frac{2}{7}\pi$, $\cos\frac{4}{7}\pi$, $\cos\frac{6}{7}\pi$ or $\cos\frac{8}{7}\pi$ see. Result (a) of above] is

$$8x^3 + 4x^2 - 4x - 1 = 0 \quad \text{...(i)}$$

Let θ denote one of the angles $\frac{2}{7}\pi, \frac{4}{7}\pi, \frac{6}{7}\pi$, then $x = \cos\theta$ is solution of (i).

Now suppose $y = \sec^2\theta$

Then $y = 1/(\cos^2\theta) = 1/x^2 \Rightarrow x^2 = 1/y \Rightarrow x = 1/\sqrt{y}$

Substituting this value of x in (i) we get

$$8\left(1/\sqrt{y}\right)^3 + 4(1/\sqrt{y})^2 - 4(1/\sqrt{y}) - 1 = 0$$

$$\left(8/y\sqrt{y}\right) + (4/y) - \left(4/\sqrt{y}\right) - 1 = 0$$

$$\Rightarrow 8 + 4/\sqrt{y} - 4y - y\sqrt{y} = 0 \quad 8 - 4y = (y-4)\sqrt{y}$$

$$\Rightarrow (8-4y)^2 = (y-4)^2\, y \Rightarrow 64 - 64y + 16)^2 = y^3 - 8y^2 + 16y$$

$\Rightarrow y^3 - 24y^2 + 80y - 64 = 0$, which is the required equation whose roots are $\sec^2\frac{2}{7}\pi$, $\sec^2\frac{4}{7}\pi$, $\sec^2\frac{6}{7}\pi$.

Example 20:

Prove that $\sin\frac{2\pi}{7}$, $\sin\frac{4\pi}{7}$ *and* $\sin\frac{8\pi}{7}$ *are the roots of the equation* $x^3 - \frac{\sqrt{7}}{2}x^2 + \frac{\sqrt{7}}{8} = 0$.

Solution :

Let θ denote any one of the angles $\frac{2}{7}\pi, \frac{4}{7}\pi$ and $\frac{8}{7}\pi$.

Then $7\theta = 2n\pi$ an even multiple of π

$\Rightarrow \quad 4\theta = 2n\pi - 3\theta \Rightarrow \sin 4\theta = \sin(2n\pi - 3\theta) = -\sin 3\theta$

$\Rightarrow \quad 2\sin 2\theta\cos 2\theta = -(3\sin\theta - 4\sin^3\theta)$

$\Rightarrow \quad 4\sin\theta\cos\theta\cos 2\theta + 3\sin\theta - 4\sin^3\theta = 0$

$\Rightarrow \quad 4\cos\theta\cos 2\theta - 4\sin^2\theta + 3 = 0$, since $\sin\theta \neq 0$

$\Rightarrow \quad 4\cos\theta(1 - 2\sin^2\theta) - 4\sin^2\theta + 3 = 0$

$\Rightarrow \quad 4\sqrt{(1-\sin^2\theta)}\ (1 - 2\sin^2\theta) - 4\sin^2\theta + 3 = 0$

$\Rightarrow \quad 4\sqrt{(1-x^2)}\ (1 - 2x^2) = 4x^2 - 3$, where $x = \sin\theta$

Squaring both sides we have $16(1 - x^2)(1 - 2x^2)^2 = (4x^2 - 3)^2$

$\Rightarrow \quad 64x^6 - 112x^4 + 56x^2 - 7 = 0$ ...(i)

which being a cubic equation in x^2 has its roots as $\sin^2 2\pi/7$, $\sin^2 4\pi/7$ and $\sin^2 8\pi/7$

$\therefore$ From (i) we have

$$\sin^2\frac{2\pi}{7} + \sin^2\frac{4\pi}{7} + \sin^2\frac{8\pi}{7} = \text{sum of the roots} = \frac{112}{64} = \frac{7}{4} \quad \text{...(ii)}$$

Also $\sin\frac{2}{7}\pi\sin\frac{4}{7}\pi + \sin\frac{4}{7}\pi\sin\frac{8}{7}\pi + \sin\frac{8}{7}\pi\sin\frac{2}{7}\pi$

$$= \frac{1}{2}\left[2\sin\frac{2}{7}\pi\sin\frac{4}{7}\pi + 2\sin\frac{8}{7}\pi\sin\frac{4}{7}\pi + 2\sin\frac{8}{7}\pi\sin\frac{2}{7}\pi\right]$$

$$= \frac{1}{2}\left[\left(\cos\frac{2}{7}\pi - \cos\frac{6}{7}\pi\right) + \left(\cos\frac{4}{7}\pi - \cos\frac{12}{7}\pi\right) + \left(\cos\frac{6}{7}\pi - \sin\frac{10}{7}\pi\right)\right]$$

$$= \frac{1}{2}\left[\cos\frac{2}{7}\pi + \cos\frac{4}{7}\pi - \cos\frac{12}{7}\pi - \cos\frac{10}{7}\pi\right]$$

$$= \frac{1}{2}\left[\cos\frac{2}{7}\pi + \cos\frac{4}{7}\pi - \cos\left(2\pi - \frac{2}{7}\pi\right) - \cos\left(2\pi - \frac{4}{7}\pi\right)\right]$$

$$= \frac{1}{2}\left[\cos\frac{2}{7}\pi + \cos\frac{4}{7}\pi - \cos\frac{2}{7}\pi - \cos\frac{4}{7}\pi\right] = 0$$

$$\Rightarrow \sin\frac{2}{7}\pi\sin\frac{4}{7}\pi + \sin\frac{4}{7}\pi\sin\frac{8}{7}\pi + \sin\frac{8}{7}\pi\sin\frac{2}{7}\pi = 0 \quad \text{...(iii)}$$

Again from equation (i) we have

$$\sin^2\frac{2}{7}\pi\sin^2\frac{4}{7}\pi\sin^2\frac{8}{7}\pi = \text{Product of the roots} = -(-7)/64 = 7/64$$

$$\Rightarrow \sin\frac{2\pi}{7}\sin\frac{4\pi}{7}\sin\frac{8\pi}{7} = -\sqrt{\frac{7}{64}} = -\frac{\sqrt{7}}{8} \quad \text{...(iv)}$$

[– ve value is taken as $\sin 8\pi/7 = -$ ve] **(Note)**

Now $\left(\sin\frac{2}{7}\pi + \sin\frac{4}{7}\pi + \sin\frac{8}{7}\pi\right)^2$

$$= \left(\sin^2 \frac{2}{7}\pi + \sin^2 \frac{4}{7}\pi + \sin^2 \frac{8}{7}\pi\right) + 2\left(\sin \frac{2}{7}\pi + \sin \frac{4}{7}\pi + \ldots\right)$$

= (7/4) + 2 (0), from (iii) and (ii)

$$\sin \frac{2}{7}\pi + \sin \frac{4}{7}\pi + \sin \frac{8}{7}\pi = \sqrt{\frac{7}{4}} = \frac{\sqrt{7}}{2} \quad \ldots(v)$$

Hence the equation whose roots are

sin 2π/7, sin 4π/7 and sin 8π/7 is

$x^3 - x^2$ (Sum of roots taken once

+ x (Sum of roots taken two at a time) – (Product of the roots) = 0

$$\Rightarrow x^3 - x^2\left(\sin \frac{2}{7}\pi + \sin \frac{4}{7}\pi + \sin \frac{8}{7}\pi\right)$$

$$+ x\left(\sin \frac{2}{7}\pi \sin \frac{4}{7}\pi + \sin \frac{4}{7}\pi \sin \frac{8}{7}\pi + \sin \frac{8}{7}\pi \sin \frac{2}{7}\pi\right)$$

$$-\left(\sin \frac{2\pi}{7} \sin \frac{4\pi}{7} \sin \frac{8\pi}{7}\right) = 0$$

$$\Rightarrow \quad x^3 - x^2\left(\sqrt{7/2}\right) + x\,(0) - \left(-\sqrt{7/8}\right) = 0$$

$$\Rightarrow x^3 - \frac{1}{2}\sqrt{7x^2} + \frac{1}{8}\sqrt{7} = 0.$$

Hence proved.

Example 21:

Form the equation whose roots are

$$\tan^2 \frac{\pi}{11}, \tan^2 \frac{2\pi}{11}, \tan^2 \frac{3\pi}{11}, \tan^2 \frac{4\pi}{11} \text{ and } \tan^2 \frac{5\pi}{11}.$$

Hence prove that

(a) $$\sec^2 \frac{\pi}{11} + \sec^2 \frac{2\pi}{11} + \sec^2 \frac{3\pi}{11} + \sec^2 \frac{4\pi}{11} + \sec^2 \frac{5\pi}{11} = 60$$

and (b) $$\cot^2 \frac{\pi}{11} + \cot^2 \frac{2\pi}{11} + \cot^2 \frac{3\pi}{11} + \cot^2 \frac{4\pi}{11} + \cot^2 \frac{5\pi}{11} = 15.$$

Solution :

Let θ be any one of the angles

$$\frac{\pi}{11}, \frac{2\pi}{11}, \frac{3\pi}{11}, \frac{4\pi}{11}, \frac{5\pi}{11}, \frac{6\pi}{11}, \frac{7\pi}{11}, \frac{8\pi}{11}, \frac{9\pi}{11}, \frac{10\pi}{11} \text{ and } \frac{11\pi}{11}$$

Then tan 11θ = 0, ∴ tan nπ = 0, for all integral values of n

$$\Rightarrow \frac{11c_1 \tan\theta - 11c_3 \tan^3\theta + 11c_5 \tan^5\theta - 11c_7 \tan^7\theta}{1 - 11c_2 \tan^2\theta + 11c_4 \tan^4\theta - 11c_6 \tan^6\theta}$$

$$\frac{+ 11c_9 \tan^9\theta - 11c_{11} \tan^{11}\theta}{+ 11c_8 \tan^8\theta - 11c_{10} \tan^{10}\theta} = 0$$

$$\Rightarrow 11c_1 \tan\theta - 11c_3 \tan^3\theta + 11c_5 \tan^5\theta - 11c_7 \tan^7\theta$$

$$+ 11c_9 \tan^9\theta - 11c_{11} \tan^{11}\theta = 0$$

$$\Rightarrow 11 \tan\theta - 165 \tan^3\theta + 462 \tan^5\theta - 330 \tan^7\theta + 55 \tan^9\theta$$

$$- \tan^{11}\theta = 0,$$

putting the values of $11c_1$, $11c_3$ etc.

$\Rightarrow 11x - 165x^3 + 462x^5 - 330x^7 + 55x^9 - x^{11} = 0$, where $x = \tan\theta$

$\Rightarrow x(x^{10} - 55x^8 + 330x^6 - 462x^4 + 165x^2 - 11) = 0$

But $x = 0$ corresponds to the value of $\theta = \dfrac{11\pi}{11} = \pi$

Hence the equation

$$x^{10} - 55x^8 + 330x^6 - 462x^4 + 165x^2 - 11 = 0 \quad ...(i)$$

has its roots as tan $\tan\dfrac{\pi}{11}, \tan\dfrac{2\pi}{11}, \tan\dfrac{3\pi}{11}, ..., \tan\dfrac{10\pi}{11}$

But this is a fifth degree equation in x^2.

$\therefore$ putting $x^2 = y$, this reduces to

$$y^5 - 55y^4 - 330y^3 - 462y^2 + 165y - 11 = 0 \quad ...(ii)$$

whose roots are

$$\tan^2\frac{\pi}{11}, \tan^2\frac{2\pi}{11}, \tan^2\frac{3\pi}{11}, \tan^2\frac{4\pi}{11} \text{ and } \tan^2\frac{5\pi}{11}$$

(a) From equation (ii) above we find that

$$\tan^2\frac{\pi}{11} + \tan^2\frac{2\pi}{11} + \tan^2\frac{3\pi}{11} + \tan^2\frac{4\pi}{11} + \tan^2\frac{5\pi}{11}$$

$$= \text{sum of the roots taken once} = \frac{55}{1} = 55$$

Adding 5 to both sides we have

$$\left(1+\tan^2\frac{\pi}{11}\right) + \left(1+\tan^2\frac{2\pi}{11}\right) + \left(1+\tan^2\frac{3\pi}{11}\right) + \left(1+\tan^2\frac{4\pi}{11}\right)$$

$$+ \left(1+\tan^2\frac{5\pi}{11}\right) = 55 + 5 \quad \textbf{(Note)}$$

$$\Rightarrow \sec^2\frac{\pi}{11} + \sec^2\frac{2\pi}{11} + \sec^2\frac{3\pi}{11} + \sec^2\frac{4\pi}{11} + \sec^2\frac{5\pi}{11} = 60.$$

Hence proved.

(b) We have proved above that the equation (ii) viz.,

$y^5 - 55y^4 + 330\,y^3 - 462y^2 + 165y - 11 = 0$ has its roots

as $\tan^2\frac{\pi}{11}, \tan^2\frac{2\pi}{11}, \tan^2\frac{3\pi}{11}, \tan^2\frac{4\pi}{11}$ and $\tan^2\frac{5\pi}{11}$

Also $\therefore \cot^2(\pi/11) = 1/\tan^2(\pi/11)$

$\therefore$ putting $x = 1/y \Rightarrow y = 1/x$ we have the equation whose roots are

$\cot^2\frac{\pi}{11}, \cot^2\frac{2\pi}{11}, \ldots, \cot^2\frac{5\pi}{11}$ as

$$\frac{1}{x^5} - \frac{55}{x^4} + \frac{330}{x^3} - \frac{462}{x^2} + \frac{165}{x} - 11 = 0$$

$$\Rightarrow 11x^5 - 165x^4 + 462x^3 - 330x^2 + 55x - 1 = 0,$$

whence $\cot^2\frac{\pi}{11} + \cot^2\frac{2\pi}{11} + \cot^2\frac{3\pi}{11} + \cot^2\frac{4\pi}{11} + \cot^2\frac{5\pi}{11}$

= sum of the roots taken once $= -(-165)/11 = 15$

Hence proved.

Example 22:

Express the following in the form $a + ib$:

(a) $\dfrac{(\cos 3\theta + i \sin 3\theta)^5}{(\cos\theta + i\sin\theta)^6}$ *(b)* $\dfrac{(\cos 3\alpha + i\sin 3\alpha)^{10}}{(\cos 2\alpha + i\sin 2\alpha)^4}$.

Solution :

(a) By De Moivre's Theorem, we have

$$\frac{(\cos 3\theta + i\sin 3\theta)^5}{(\cos\theta + i\sin\theta)^6} = \frac{\cos(5\times 3\theta) + i\sin(5\times 3\theta)}{\cos 6\theta + i\sin 6\theta}$$

$$= \frac{\cos 15\theta + i\sin 15\theta}{\cos 6\theta + i\sin 6\theta}$$

$= (\cos 15\theta + i\sin 15\theta)(\cos 6\theta - i\sin 6\theta)$ [Remark 1]

$= \cos(15\theta - 6\theta) + i\sin(15\theta - 6\theta)$ [Remark 3]

$= \cos 9\theta + i \sin 9\theta.$

(b) $\dfrac{(\cos\alpha + i\sin\alpha)^{10}}{(\cos 2\alpha + i\sin 2\alpha)^4} = \dfrac{\cos 10\alpha + i\sin 10\alpha}{\cos 8\alpha + i\sin 8\alpha}$

$= \cos(10\alpha - 8\alpha) + i\sin(10\alpha - 8\alpha)$

$= \cos 2\alpha + i\sin 2\alpha.$

Example 23:

Find the modulus and the argument of the complex number

$$z = \frac{(\sin\alpha + i\ coa\ \alpha)^4}{(\cos\alpha - i\sin\alpha)^3}.$$

Solution :

$(\sin\alpha + i\cos\alpha)^4 = \{\cos(\pi/2 - \alpha) + i\sin(\pi/2 - \alpha)\}^4$

$= \cos(2\pi - 4\alpha) + i\sin(2\pi - 4\alpha)$ (De Moivre's Theorem)

Also $(\cos\alpha - i\sin\alpha)^3 = \cos 3\alpha - i\sin 3\alpha.$

$\therefore z = \{\cos(2\pi - 4\alpha) + i\sin(2\pi - 4\alpha)\}\ \{\cos 3\alpha + i\sin 3\alpha\}$

$= \cos\{(2\pi - 4\alpha) + 3\alpha\} + i\sin\{(2\pi - 4\alpha) + 3\alpha\}$

$= \cos(2\pi - \alpha) + i\sin(2\pi - \alpha)$

$= \cos\alpha - i\sin\alpha = \cos(-\alpha) + i\sin(-\alpha).$

Hence the modulus of z is $\sqrt{\cos^2\alpha + \sin^2\alpha = 1}$ and the argument of z is $-\alpha$.

Example 24:

Find the modulus and the argument of the complex number $\dfrac{(1 + \cos\theta + i\sin\theta)^5}{(\cos\theta + i\sin\theta)^3}$.

Solution :

Let $z = \dfrac{(1 + \cos\theta + i\sin\theta)^5}{(\cos\theta + i\sin\theta)^3}$

$= \dfrac{\{1 + 2\cos^2(\theta/2) - 1 + 2i\sin\theta/2\cos\theta/2\}^5}{\cos 3\theta + i\sin 3\theta}$

$= \dfrac{32\cos^5(\theta/2)\{\cos\theta/2 + i\sin\theta/2\}^5}{\cos 3\theta + i\sin 3\theta}$

$= 32 \cos^5 (\theta/2)\{\cos (5\theta/2) + i \sin (5\theta/2)\} (\cos 3\theta - i \sin 3\theta)$

$= 32 \cos^5 (\theta/2) [\cos \{(5\theta/2) - 3\theta\} + i \sin \{(5\theta/2) - 3\theta\}]$

$= 32 \cos^5 (\theta/2) [\cos (-\theta/2) + i \sin (-\theta/2)]$.

Hence, the modulus and argument of z is $32 |\cos^5 (\theta/2)|$ and $(-\theta/2)$ respectively.

Example 25:

if $z = \cos\theta + i \sin\theta$, show that

$$z^n + \frac{1}{z^n} = 2 \cos n\theta, \quad z^n - \frac{1}{z^n} = 2i \sin n\theta.$$

Solution :

We have $\frac{1}{z} = \frac{1}{\cos\theta + i\sin\theta} = \cos\theta - i\sin\theta$.

$\therefore z^n = (\cos\theta + i \sin\theta)^n = \cos n\theta + i \sin n\theta$,

and $\frac{1}{z^n} (\cos\theta - i \sin\theta)^n = \cos n\theta - i \sin n\theta$.

Hence $z^n + \frac{1}{z^n} = 2 \cos n\theta$ and $z^n - \frac{1}{z^n} = 2i \sin n\theta$.

Example 26:

If $z = \cos\theta + i \sin\theta$, show that

$$\frac{z^{2n} - 1}{z^{2n} + 1} = i \tan n\theta, \text{ n being an integer.}$$

Solution :

We have

$$\frac{z^{2n} - 1}{z^{2n} + 1} = \frac{(\cos\theta + i\sin\theta)^{2n} - 1}{(\cos\theta + i\sin\theta)^{2n} - 1}$$

$$= \frac{\cos 2n\theta + i \sin 2n\theta - 1}{\cos 2n\theta + i \sin 2n\theta + 1} \text{ (De Moivre's Theorem)}$$

$$= \frac{(1 - 2\sin^2 n\theta) + 2i \sin n\theta \cos n\theta - 1}{(2\cos^2 n\theta - 1) + 2i \sin n\theta \cos n\theta + 1}$$

$$= \frac{i \sin n\theta \cos n\theta + i^2 \sin^2 n\theta}{\cos^2 n\theta + i \sin n\theta \cos n\theta} \qquad (\because i^2 = -1)$$

$$= \frac{i \sin n\theta (\cos n\theta + i \sin n\theta)}{\cos n\theta (\cos n\theta + i \sin n\theta)} = i \tan n\theta.$$

Example 27:

If n is an integer, show that

$$(1 + \cos\theta + i\sin\theta)^n + (1 + \cos\theta - i\sin\theta)^n = 2^{n+1}\cos^n\frac{\theta}{2}\cos\frac{n\theta}{2}.$$

Solution :

$$(1 + \cos\theta + i\sin\theta)^n + (1 + \cos\theta - i\sin\theta)^n$$

$$= [1 + 2\cos^2\theta/2 - 1 + 2i\sin\theta/2\cos\theta/2]^n + [1 + 2\cos^2\theta/2 - 1 - 2i\sin\theta/2\cos\theta/2]^n$$

$$= 2^n\cos^n\theta/2\,[\cos\theta/2 + i\sin\theta/2]^n + 2^n\cos^n\theta/2\,[\cos\theta/2 - i\sin\theta/2]^n$$

$$= 2^n\cos^n\frac{\theta}{2}\left[\left(\cos\frac{n\theta}{2} + i\sin\frac{n\theta}{2}\right) + \left(\cos\frac{n\theta}{2} - i\sin\frac{n\theta}{2}\right)\right]$$

[By De Moivre's Theorem]

$$= 2^n\cos^n\frac{\theta}{2}.2\cos\frac{n\theta}{2} = 2^{n+1}\cos^n\frac{\theta}{2}\cos\frac{n\theta}{2}.$$

Example 28:

Show that

(i) $(\sin\theta + i\cos\theta)^n = \cos n\left(\frac{\pi}{2} - \theta\right) + i\sin n\left(\frac{\pi}{2} - \theta\right)$,

n being an integer.

(ii) $\dfrac{1 + \cos\theta + i\sin\theta}{1 + \cos\theta - i\sin\theta} = \cos\theta + i\sin\theta.$

Solution :

We have

(i) $(\sin\theta + i\cos\theta)^n = \left[\cos\left(\frac{\pi}{2} - \theta\right) + i\sin\left(\frac{\pi}{2} - \theta\right)\right]^n$

$$= \cos n\left(\frac{\pi}{2} - \theta\right) + i\sin n\left(\frac{\pi}{2} - \theta\right).$$

(ii) $\dfrac{1 + \cos\theta + i\sin\theta}{1 + \cos\theta - i\sin\theta} = \dfrac{1 + 2\cos^2\theta/2 - 1 + 2i\sin\theta/2\cos\theta/2}{1 + 2\cos^2\theta/2 - 1 - 2i\sin\theta/2\cos\theta/2}$

$$= \frac{\cos\theta/2\,[\cos\theta/2 + i\sin\theta/2]}{\cos\theta/2\,[\cos\theta/2 - i\sin\theta/2]}$$

$$= \frac{(\cos\theta/2 + \sin\theta/2)(\cos\theta/2 + i\sin\theta/2)}{(\cos\theta/2 - i\sin\theta/2)(\cos\theta/2 + i\sin\theta/2)}$$

$= (\cos\theta/2 + i\sin\theta/2)^2 = \cos\theta + i\sin\theta$, (by De Moivre's Theorem)

Example 29:

We have

$$\left[\frac{1+\sin\phi + i\cos\phi}{1+\sin\phi - i\cos\phi}\right] = \cos\left(\frac{n\pi}{2} - n\phi\right) + i\sin\left(\frac{n\pi}{2} - n\phi\right).$$

Solution :

We have

$$\left[\frac{1+\sin\phi + i\cos\phi}{1+\sin\phi - i\cos\phi}\right] = \left[\frac{1+\cos\left(\frac{\pi}{2}-\phi\right) + i\sin\left(\frac{\pi}{2}-\phi\right)}{1+\cos\left(\frac{\pi}{2}-\phi\right) - \sin\left(\frac{\pi}{2}-\phi\right)}\right]$$

$$= \left[\frac{1+2\cos^2\left(\frac{\pi}{4}-\frac{\phi}{2}\right) - 1 + 2i\sin\left(\frac{\pi}{4}-\frac{\phi}{2}\right)\cos\left(\frac{\pi}{4}-\frac{\phi}{2}\right)}{1+2\cos^2\left(\frac{\pi}{4}-\frac{\phi}{2}\right) - 1 - 2i\sin\left(\frac{\pi}{4}-\frac{\phi}{2}\right)\cos\left(\frac{\pi}{4}-\frac{\phi}{2}\right)}\right]^n$$

$$= \frac{2^n\cos^n\left(\frac{\pi}{4}-\frac{\phi}{2}\right)\left\{\cos\left(\frac{\pi}{4}-\frac{\phi}{2}\right) + i\sin\left(\frac{\pi}{4}-\frac{\phi}{2}\right)\right\}^n}{2^n\cos^n\left(\frac{\pi}{4}-\frac{\phi}{2}\right)\left\{\cos\left(\frac{\pi}{4}-\frac{\phi}{2}\right) - i\sin\left(\frac{\pi}{4}-\frac{\phi}{2}\right)\right\}^n}$$

$$= \frac{\cos n\left(\frac{\pi}{4}-\frac{\phi}{2}\right) + i\sin n\left(\frac{\pi}{4}-\frac{\phi}{2}\right)}{\cos n\left(\frac{\pi}{4}-\frac{\phi}{2}\right) - i\sin n\left(\frac{\pi}{4}-\frac{\phi}{2}\right)} \quad \text{(De Moivre's Theorem)}$$

$$= \left[\cos n\left(\frac{\pi}{4}-\frac{\phi}{2}\right) + i\sin n\left(\frac{\pi}{4}-\frac{\phi}{2}\right)\right] \times \left[\cos n\left(\frac{\pi}{4}-\frac{\phi}{2}\right) + i\sin n\left(\frac{\pi}{4}-\frac{\phi}{2}\right)\right]$$

$$= \left\{\cos n\left(\frac{\pi}{4}-\frac{\phi}{2}\right) + i\sin n\left(\frac{\pi}{4}-\frac{\phi}{2}\right)\right\}^2$$

$$= \cos\left(\frac{n\pi}{2} - n\phi\right) + i \sin\left(\frac{n\pi}{2} - n\phi\right). \text{ [De Moivre's Theorem]}$$

Example 30:

Express the following in the form x + iy

$$\left[\frac{1 + \sin\theta + i\cos\theta}{1 + \sin\theta - i\cos\theta}\right]^n.$$

Solution :

It is same as Example where

$$x + iy = \cos\left(\frac{n\pi}{2} - n\theta\right) + i \sin\left(\frac{n\pi}{2} - n\theta\right).$$

Example 31:

Find the modulus and argument of

$$\left(\frac{1 + \cos\theta + i\sin\theta}{1 + \cos\theta - i\sin\theta}\right)^n,$$

n is a positive integer and $\theta \neq (2k + 1)\pi$.

Solution :

$$\text{Let } z = \frac{1 + \cos\theta + i\sin\theta}{1 + \cos\theta - i\sin\theta}$$

$$= \frac{1 + 2\cos^2(\theta/2) - 1 + 2i\sin\theta/2\cos\theta/2}{1 + 2\cos^2(\theta/2) - 1 - 2i\sin\theta/2\cos\theta/2}$$

$$= \frac{\cos\theta/2 + i\sin\theta/2}{\cos\theta/2 - i\sin\theta/2}, \text{ as } \cos\theta/2 \neq 0$$

$$= (\cos\theta/2 + i\sin\theta/2)(\cos\theta/2 + i\sin\theta/2)$$

$$= (\cos\theta/2 + i\sin\theta/2)^2$$

$$\therefore z^n = (\cos\theta/2 + i\sin\theta/2)^{2n} = \cos n\theta + i\sin n\theta$$

(De Moivre's Theorem)

Hence the modulus of z^n is $\sqrt{\cos^2 n\theta + \sin^2 n\theta} = 1$ and the argument of z^n is $n\theta$.

Example 32:

Show that

$\{(\cos\alpha + i\sin\alpha) + (\cos\beta + i\sin\beta)\}^n + \{(\cos\alpha - i\sin\alpha) + (\cos\beta - i\sin\beta)\}^n$

$= 2^{n+1}\cos^n\left(\dfrac{\alpha-\beta}{2}\right)\cos\left\{\dfrac{n}{2}(\alpha+\beta)\right\}$, *n being an integer.*

Solution :

L.H.S. $= \{(\cos\alpha + \cos\beta) + i(\sin\alpha + \sin\beta)\}^n + \{(\cos\alpha + \cos\beta) - i(\sin\alpha + \sin\beta)\}^n$

$$= \left\{2\cos\frac{\alpha+\beta}{2}\cos\frac{\alpha-\beta}{2} + 2i\sin\frac{\alpha+\beta}{2}\cos\frac{\alpha-\beta}{2}\right\}^n + \left\{2\cos\frac{\alpha+\beta}{2}\cos\frac{\alpha-\beta}{2} - 2i\sin\frac{\alpha+\beta}{2}\cos\frac{\alpha-\beta}{2}\right\}^n$$

$$= 2^n\cos^n\left(\frac{\alpha-\beta}{2}\right)\left\{\cos\frac{\alpha+\beta}{2} + i\sin\frac{\alpha+\beta}{2}\right\}^n + 2^n\cos^n\left(\frac{\alpha-\beta}{2}\right)\left\{\cos\frac{\alpha+\beta}{2} - i\sin\frac{\alpha+\beta}{2}\right\}^n$$

$$= 2^n\cos^n\left(\frac{\alpha-\beta}{2}\right)\left[\left\{\cos\frac{n(\alpha+\beta)}{2} + i\sin\frac{n(\alpha+\beta)}{2}\right\} + \left\{\cos\frac{n(\alpha+\beta)}{2} - i\sin\frac{n(\alpha+\beta)}{2}\right\}\right]$$ (by De Moivre's Theorem)

$$= 2^n\cos^n\left(\frac{\alpha-\beta}{2}\right)\left[2\cos\frac{n(\alpha+\beta)}{2}\right]$$

$$= 2^{n+1}\cos^n\left(\frac{\alpha-\beta}{2}\right)\cos\left\{\frac{n}{2}(\alpha+\beta)\right\}$$

Example 33:

If $m = \cos\alpha + i\sin\alpha$ *and* $n = \cos\beta + i\sin\beta$, *prove that*

$$\frac{m-n}{m+n} = i\tan\left(\frac{\alpha-\beta}{2}\right).$$

Solution :

We have

$m - n = (\cos\alpha - \cos\beta) + i(\sin\alpha - \sin\beta)$.

$$= -2 \sin \frac{\alpha+\beta}{2} \sin \frac{\alpha-\beta}{2} + 2i \cos \frac{\alpha+\beta}{2} \sin \frac{\alpha-\beta}{2}$$

$$\therefore m - n = 2i \sin \frac{\alpha-\beta}{2}\left[\cos \frac{\alpha+\beta}{2} + i \sin \frac{\alpha+\beta}{2}\right]$$

Now $m + n = (\cos \alpha + \cos \beta) + i (\sin \alpha + \sin \beta)$ ...(1)

$$= 2 \cos \frac{\alpha+\beta}{2} \cos \frac{\alpha-\beta}{2} + 2i \sin \frac{\alpha+\beta}{2} \cos \frac{\alpha-\beta}{2}$$

$$\therefore m + n = 2 \cos \frac{\alpha-\beta}{2}\left[\cos \frac{\alpha+\beta}{2} + i \sin \frac{\alpha+\beta}{2}\right]. \quad ...(2)$$

From (1) and (2), we obtain

$$\frac{m+n}{m+n} = i \tan\left(\frac{\alpha-\beta}{2}\right).$$

Example 34:

If $x_r = \cos \frac{\pi}{2^r} + i \sin \frac{\pi}{2^r}$, *prove that*

$x_1 x_2 x_3 \ldots$ *to* $\infty = -1$.

Solution :

We have

$$x_1x_2x_2\ldots \infty = \left(\cos \frac{\pi}{2} + i \sin \frac{\pi}{2}\right)\left(\cos \frac{\pi}{2^2} + i \sin \frac{\pi}{2^2}\right)\left(\cos \frac{\pi}{2^3} + i \sin \frac{\pi}{2^3}\right)\ldots\ldots$$

$$= \cos\left(\frac{\pi}{2}+\frac{\pi}{2^3}+\frac{\pi}{2^3}+\ldots\ldots\right) + i \sin\left(\frac{\pi}{2}+\frac{\pi}{2^3}+\frac{\pi}{2^3}+\ldots\ldots\right)$$

[Remark 4, p. 2] ...(1)

We note that $\frac{\pi}{2}+\frac{\pi}{2^2}+\frac{\pi}{2^3}+\ldots\ldots$ is a G.P., whose common ratio is r = 1/2 and the first term is $\alpha = \pi/2$. Thus

$$\frac{\pi}{2}+\frac{\pi}{2^2}+\frac{\pi}{2^3}+\ldots\ldots = \frac{a}{1-r} = \frac{\pi/2}{1-\frac{1}{2}} = \pi.$$

From (1), $x_1x_2x_3\ldots \infty = \cos \pi + i \sin \pi = -1$.

Example 35:

If $\sin\theta + \sin\phi + \sin\psi = 0 = \cos\theta + \cos\phi + \cos\psi$, show that $\sin 3\theta + \sin 3\phi + \sin 3\psi = 3\sin(\theta + \phi + \psi)$.

and $\cos 3\theta + \cos 3\phi + \cos 3\psi = 3\cos(\theta + \phi + \psi)$.

Solution :

Let $x = \cos\theta + i\sin\theta$, $y = \cos\phi + i\sin\phi$, $z = \cos\psi + i\sin\psi$.

Now $x + y + z = (\cos\theta + \cos\phi + \cos\psi) + i(\sin\theta + \sin\phi + \sin\psi)$

$\Rightarrow x + y + z = 0$

$\Rightarrow x^3 + y^3 + z^3 = 3xyz.$

Thus $(\cos\theta + i\sin\theta)^3 + (\cos\phi + i\sin\phi)^3 + (\cos\psi + i\sin\psi)^3$

$= 3(\cos\theta + i\sin\theta)(\cos\phi + i\sin\phi)(\cos\psi + i\sin\psi)$

$\Rightarrow (\cos 3\theta + i\sin 3\theta) + (\cos 3\phi + i\sin 3\phi) + (\cos 3\psi + i\sin 3\psi)$

$= 3\{\cos(\theta + \phi + \psi) + i\sin(\theta + \phi + \psi)\}$. [Remark 4]

Comparing real and imaginary parts, we get

$\cos 3\theta + \cos 3\phi + \cos 3\psi = 3\cos(\theta + \phi + \psi)$,

and $\sin 3\theta + \sin 3\phi + \sin 3\psi = 3\sin(\theta + \phi + \psi)$.

Example 36:

If $\cos\alpha + 2\cos\beta + 3\cos\gamma = \sin\alpha + 2\sin\beta + 3\sin\gamma = 0$, *prove that*

$$\cos 3\alpha + 8\cos 3\beta + 27\cos 3\gamma = 18\cos(\alpha + \beta + \gamma),$$
$$\sin 3\alpha + 8\sin 3\beta + 27\sin 3\gamma = 18\sin(\alpha + \beta + \gamma).$$

Solution :

Let $a = \cos\alpha + i\sin\alpha$, $b = 2\cos\beta + 2i\sin\beta$

and $c = 3\cos\gamma + 3i\sin\gamma$. Then

$a + b + c = (\cos\alpha + 2\cos\beta + 3\cos\gamma) +$
$i(\sin\alpha + 2\sin\beta + 3\sin\gamma) = 0$

We know $a + b + c = 0 \Rightarrow a^3 + b^3 + c^3 = 3abc$

$\Rightarrow (\cos\alpha + i\sin\alpha)^3 + (2\cos\beta + 2i\sin\beta)^3 + (3\cos\gamma + 3i\sin\gamma)^3$

$= 3(\cos\alpha + i\sin\alpha)(2\cos\beta + 2i\sin\beta)(3\cos\gamma + 3i\cos\gamma)$

$\Rightarrow (\cos 3\alpha + i\sin 3\alpha) + 8(\cos 3\beta + i\sin 3\beta) + 27(\cos 3\gamma + i\sin 3\gamma)$

$= 18(\cos\alpha + i\sin\alpha)(\cos\beta + i\sin\beta)(\cos\gamma + i\sin\gamma)$

$\Rightarrow (\cos 3\alpha + 8\cos 3\beta + 27\cos 3\gamma) + i(\sin 3\alpha + 8\sin 3\beta + 27\sin 3\gamma)$

$= 18[\cos(\alpha + \beta + \gamma) + i\sin(\alpha + \beta + \gamma)]$.

Comparing real and imaginary parts, we get
$\cos 3\alpha + 8 \cos 3\beta + 27 \cos 3\gamma = 18 \cos (\alpha + \beta + \gamma)$,
and $\sin 3\alpha + 8 \sin 3\beta + 27 \sin 3\gamma = 18 \sin (\alpha + \beta + \gamma)$.

Example 37:

If $a = \cos \alpha + i \sin \alpha$, $b = \cos \beta + i \sin \beta$,

$c = \cos \gamma + i \sin \gamma$, *and* $\frac{a}{b}+\frac{b}{c}+\frac{c}{a}=-1$, *prove that*

$\cos (\beta - \gamma) + \cos (\gamma - \alpha) + \cos (\alpha - \beta) + 1 = 0.$

Solution :

We have $1/a = \cos \alpha - i \sin \alpha$, $1/b = \cos \beta - i \sin \beta$.

Now $a/b = (\cos \alpha + i \sin \alpha) (\cos \beta - i \sin \beta)$

$\Rightarrow a/b = \cos (\alpha - \beta) + i \sin (\alpha - \beta)$.

Similarly $b/c = \cos (\beta - \gamma) + i \sin (\beta - \gamma)$,

and $c/a = \cos (\gamma - \alpha) + i \sin (\gamma - \alpha)$.

Putting these values in $\frac{a}{b}+\frac{b}{c}+\frac{c}{a}=-1$, we get

$$[\cos (\alpha - \beta) + \cos (\beta - \gamma) + \cos (\gamma - \alpha)] + i [\sin (\alpha - \beta) + \sin (\beta - \gamma) + \sin (\gamma - \alpha)]$$
$$= -1 = -1 + 0\, i.$$

Comparing real and imaginary parts, we get

$\cos (\alpha - \beta) + \cos (\beta - \gamma) + \cos (\gamma - \alpha) = -1.$

Hence $\cos (\alpha - \beta) + \cos (\beta - \gamma) + \cos (\gamma - \alpha) + 1 = 0.$

Example 38:

If $a \cos \alpha + i \sin \alpha$, $b = \cos \beta + i \sin \beta$,

$c = \cos \gamma + i \sin \alpha$, *prove that* $(b + c) (c + a) (a + b)/abc$

$= 8 \cos \frac{1}{2} (\beta - \gamma) \cos \frac{1}{2} (\gamma - \alpha) \cos \frac{1}{2} (\alpha - \beta)$.

Solution :

$$\frac{(b+c)(c+a)(a+b)}{abc} = \left(\frac{b+c}{c}\right)\left(\frac{c+a}{a}\right)\left(\frac{a+b}{b}\right)$$

$$= \left(1+\frac{a}{b}\right)\left(1+\frac{b}{c}\right)\left(1+\frac{c}{a}\right). \quad ...(1)$$

By Example 16, $a/b = \cos(\alpha - \beta) + i \sin(\alpha - \beta)$

$$\therefore \left(1+\frac{a}{b}\right) = 1+2\cos^2\frac{1}{2}(\alpha-\beta) - 1 + 2i\sin\frac{1}{2}(\alpha-\beta)\cos\frac{1}{2}(\alpha-\beta)$$

$$= 2\cos\frac{1}{2}(\alpha-\beta)\left\{\cos\frac{1}{2}(\alpha-\beta) + i\sin\frac{1}{2}(\alpha-\beta)\right\} \text{ etc.} \quad ...(2)$$

From (1) and (2), we obtain

$$\frac{(b+c)(c+a)(a+b)}{abc} = 8\cos\frac{1}{2}(\alpha-\beta)\cos\frac{1}{2}(\beta-\gamma)\cos\frac{1}{2}(\gamma-\alpha)$$

$$\times\left[\cos\left\{\frac{1}{2}(\alpha-\beta)+\frac{1}{2}(\beta-\gamma)+\frac{1}{2}(\gamma-\alpha)\right\}\right.$$

$$\left.+\sin\left\{\frac{1}{2}(\alpha-\beta)+\frac{1}{2}(\beta-\gamma)+\frac{1}{2}(\gamma-\alpha)\right\}\right]$$

$$= 8\cos\frac{1}{2}(\alpha-\beta)\cos\frac{1}{2}(\beta-\gamma)\cos\frac{1}{2}(\gamma-\alpha).$$

$(\because \cos 0 + i \sin 0 = 1)$

Example 39:

If $a = \cos 2\alpha + i \sin 2\alpha$, $b = \cos 2\beta + \sin 2\beta$, $c = \cos 2\gamma + i \sin 2\gamma$, $d = \cos 2\delta + i \sin 2\delta$, *show that*

$$\sqrt{abcd} + \frac{1}{\sqrt{abcd}} = 2\cos(\alpha+\beta+\gamma+\delta).$$

Solution :

We have

$$abcd = \cos(2\alpha+2\beta+2\gamma+2\delta) + i\sin(2\alpha+2\beta+2\gamma+2\delta)$$

$$\therefore \sqrt{abcd} = [\cos(2\alpha+2\beta+2\gamma+2\delta) + i\sin(2\alpha+2\beta+2\gamma+2\delta)]^{1/2}$$

$$\Rightarrow \sqrt{abcd} = \cos(\alpha+\beta+\gamma+\delta) + i\sin(\alpha+\beta+\gamma+\delta). \quad ...(1)$$

[De Moivre's Theorem]

$$\therefore \frac{1}{\sqrt{abcd}} = \cos(\alpha+\beta+\gamma+\delta) - i\sin(\alpha+\beta+\gamma+\delta). \quad ...(2)$$

Adding (1) and (2), we obtain

$$\sqrt{abcd} + \frac{1}{\sqrt{abcd}} = 2\cos(\alpha+\beta+\gamma+\delta).$$

Example 41:

If $\cos\alpha+\cos\beta+\cos\gamma=\sin\alpha+\sin\beta+\sin\gamma=0.$

then $\Sigma\cos 4\alpha=2\Sigma\cos 2\alpha(\beta+\gamma)$, $\Sigma\sin 4\alpha=2\Sigma\sin 2(\beta+\gamma)$.

Solution :

Let $a=\cos\alpha+i\sin\alpha$, $b=\cos\beta+i\sin\beta$

$c=\cos\gamma+i\sin\gamma$.

Then $a+b+c=0\Rightarrow a+b=-c\Rightarrow(a+b)^2=c^2$

$\Rightarrow a^2+b^2-c^2=-2ab\Rightarrow(a^2+b^2-c^2)^2=4a^2b^2$

$\Rightarrow a^4+b^4+c^4=2(a^2b^2+b^2c^2+c^2a^2)$. ...(1)

Now $a^2=\cos 2\alpha+i\sin 2\alpha$ and $a^4=\cos 4\alpha+i\sin 4\alpha$ etc.

We have $a^4+b^4+c^4=(\cos 4\alpha+i\sin 4\alpha)+(\cos 4\beta+i\sin 4\beta)$

$+(\cos 4\gamma+i\sin 4\gamma)$

$=\Sigma\cos 4\alpha+i\,S\sin 4\alpha$. ...(2)

Now $2(b^2c^2+c^2a^2+a^2b^2)=2(\cos 2\beta+i\sin 2\beta)(\cos 2\gamma+i\sin 2\gamma)$

$+......+.......$

$=2[\cos 2(\beta+\gamma)+i\sin 2(\beta+\gamma)]+......+.....$

$=2\Sigma\cos 2(\beta+\gamma)+i\,2\Sigma\sin 2(\beta+\gamma)$. ...(3)

Using (2) and (3) in (1), we obtain

$\Sigma\cos 4\alpha+i\Sigma\sin 4\alpha=2\Sigma\cos 2(\beta+\gamma)+i\,2\Sigma\sin 2(\beta+\gamma)$.

Comparing real and imaginary parts, we get

$\Sigma\cos 4\alpha=2\Sigma\cos 2(\beta+\gamma)$,

and $\Sigma\sin 4\alpha=2\Sigma\sin 2(\beta+\gamma)$.

Example 42:

If $a=\cos\alpha+i\sin\alpha$, $b=\cos\beta+i\sin\beta$,

$c=\cos\gamma+i\sin\gamma$ *and* $d=\cos\delta+i\sin\delta$; *prove that*

(i) $(a+b)(c+d)$

$$=4\cos\frac{\alpha-\beta}{2}\cos\frac{\gamma-\delta}{2}\left[\cos\frac{\alpha+\beta+\gamma+\delta}{2}+i\sin\frac{\alpha+\beta+\gamma+\delta}{2}\right]$$

(ii) $ab+cd$

$$=2\cos\frac{\alpha+\beta-\gamma-\delta}{2}\left[\cos\frac{\alpha+\beta+\gamma+\delta}{2}+i\sin\frac{\alpha+\beta+\gamma+\delta}{2}\right]$$

Solution :

(i) $a + b = (\cos\alpha + \cos\beta) + i(\sin\alpha + \sin\beta)$

$$\Rightarrow a + b = 2\cos\frac{\alpha+\beta}{2}\cos\frac{\alpha-\beta}{2} + 2i\sin\frac{\alpha+\beta}{2}\cos\frac{\alpha-\beta}{2}$$

$$c + d = 2\cos\frac{\gamma+\delta}{2}\cos\frac{\gamma-\delta}{2} + 2i\sin\frac{\gamma+\delta}{2}\cos\frac{\gamma-\delta}{2}.$$

$$\text{Now } (a+b)(c+d) = 4\cos\frac{\alpha-\beta}{2}\cos\frac{\gamma-\delta}{2}\left\{\cos\frac{\alpha+\beta}{2} + i\sin\frac{\gamma+\beta}{2}\right\}$$

$$\times\left\{\cos\frac{\gamma+\delta}{2} + i\sin\frac{\gamma+\delta}{2}\right\}$$

$$= 4\cos\frac{\alpha-\beta}{2}\cos\frac{\gamma-\delta}{2}\left[\cos\left(\frac{\alpha+\beta}{2}+\frac{\gamma+\delta}{2}\right)\right.$$

$$\left. + i\sin\left(\frac{\alpha+\beta}{2}+\frac{\gamma+\delta}{2}\right)\right].$$

(iii) $ab + cd = \{\cos(\alpha+\beta) + i\sin(\alpha+\beta)\} + \{\cos(\gamma+\delta) + i\sin(\gamma+\delta)\}$

$$= \{\cos(\alpha+\beta) + \cos(\gamma+\delta)\} + i\sin(\alpha+\beta) + \sin(\gamma+\delta)\}$$

$$= 2\cos\frac{\alpha+\beta+\gamma+\delta}{2}\cos\frac{\alpha+\beta-\gamma-\delta}{2}$$

$$+ 2i\sin\frac{\alpha+\beta+\gamma+\delta}{2}\cos\frac{\alpha+\beta-\gamma-\delta}{2}$$

$$= 2\cos\frac{\alpha+\beta-\gamma-\delta}{2}\left[\cos\frac{\alpha+\beta-\gamma-\delta}{2} + i\sin\frac{\alpha+\beta+\gamma+\delta}{2}\right]$$

Example 43:

If $z = \cos 2\theta + i\sin 2\theta$, $\omega = \cos 2\phi + i\sin 2\phi$; show that

(a) $z^m w^n + \dfrac{1}{z^m\omega^n} = 2\cos 2(m\theta + n\phi)$,

(b) $\dfrac{z^m}{\omega^n} + \dfrac{\omega^n}{z^m} = 2\cos 2(m\theta - n\phi)$, *m and n being integers.*

Solution :

We have $z^m = (\cos 2\theta + i\sin 2\theta)^m$

$= \cos 2m\theta + i\sin 2m\theta,$

and $w^n = (\cos 2\phi + i\sin 2\phi)^n = \cos 2n\phi + i\sin 2n\phi.$

Since $\frac{1}{z}$ = cos 2θ – i sin 2θ, $\frac{1}{z^m}$ = cos 2mθ – i sin 2mθ.

Similarly $\frac{1}{\omega^n}$ = cos 2nϕ – i sin 2nϕ.

(a) $z^m w^n + \frac{1}{z^m \omega^n}$ = (cos 2 mθ + i sin 2mθ) (cos 2nϕ + i sin 2nϕ)

+ {cos 2nϕ – i sin 2nθ} {cos 2nϕ – i sin 2nϕ}

= [cos 2 (nθ + nϕ) + i sin 2 (mθ + nϕ)

+ [cos 2 (mθ + nϕ) – i sin 2 (mθ + mϕ)]

= 2 cos 2 (mθ + nϕ).

(b) $\frac{z^m}{\omega^n} + \frac{\omega^n}{z^m}$ = (cos 2mθ + i sin 2mθ) (cos 2nϕ – i sin 2nϕ)

+ (cos 2nϕ + i sin 2nϕ) (cos 2mθ – i sin 2mθ)

= [cos 2 (mθ – nϕ) + i sin 2 (mθ – nϕ)]

+ [cos 2 (mθ – nϕ) – i sin 2 (mθ – nϕ)]

= 2 cos 2 (mθ – nϕ).

Example 44:

By expanding (1 + x)ⁿ by Binomial Theorem and then putting x = cos 2θ + i sin 2θ + i sin 2θ, prove that

(i) $2^n \cos^n \theta \cos n\theta = \sum_{r=0}^{n} {}^nC_r \cos 2r\theta,$

and (ii) $2^n \theta \sin n\theta = \sum_{r=0}^{n} {}^nC_r \sin 2r\theta.$

Solution :

By Binomial Theorem,

$(1 + x)^n = {}^nC_0 + {}^nC_1 x + {}^nC_2 x^2 + ... + {}^nC_n x^n$

i.e., $(1 + x)^n \sum_{r=0}^{n} {}^nC_r x^r$

Putting x = cos 2θ + i sin 2θ, we get

$(1 + \cos 2\theta + i \sin 2\theta)^n = \sum_{r=0}^{n} {}^nC_r (\cos 2\theta + i \sin 2\theta)^r$

$\Rightarrow (2 \cos^2 \theta + i\, 2 \sin \theta \cos \theta)^n = \sum_{r=0}^{n} {}^nC_r (\cos 2r\theta + i \sin 2r\theta)$

$$\Rightarrow 2^n \cos^n \theta\, [\cos \theta + i \sin \theta]^n = \sum_{r=0}^{n} {}^nC_r \cos 2r\theta + i \sum_{r=0}^{n} {}^nC_r \sin 2r\theta$$

$$\Rightarrow 2^n \cos^n \theta\, (\cos n\theta + i \sin n\theta) = \sum_{r=0}^{n} {}^nC_r \cos 2r\theta + i \sum_{r=0}^{n} {}^nC_r \sin 2r\theta.$$

Equating real and imaginary parts, we get

$$2^n \cos^n \theta \cos n\theta = \sum_{r=0}^{n} {}^nC_r \cos 2r\theta,$$

and $$2^n \cos^n \theta \sin n\theta = \sum_{r=0}^{n} {}^nC_r \sin 2r\theta.$$

Example 45:

If l, m, n,.... are positive integers and

$2 \cos \theta = a + 1/a$, $2 \cos \phi = b + 1/b$, etc. show that

$$a^l b^m c^n ... + \frac{1}{a^l b^m c^n ...} = 2 \cos (l\theta + m\phi + ...).$$

Solution :

$$2 \cos \theta = a + \frac{1}{a} \Rightarrow a^2 - 2a \cos \theta + 1 = 0.$$

$$\therefore a = \frac{2 \cos \theta + \sqrt{4 \cos^2 \theta - 4}}{2} = \cos \theta \pm i \sin \theta.$$

We may take $a = \cos \theta + i \sin \theta$.

Now $a^l = (\cos \theta + i \sin \theta)^l = \cos l\theta + i \sin l\theta$.

Similarly $b^m = (\cos \phi + i \sin \phi) = \text{cow } m\phi + i \sin m\phi$ etc.

Thus $a^l b^m c^n .... = (\cos l\theta + i \sin l\theta) (\cos m\phi + i \sin m\phi)$.....

$\Rightarrow a^l b^m c^n .... = \cos (l\theta + m\phi + ...) + i \sin (l\theta + m\phi + ...)$.

$$\therefore \frac{1}{a^l b^m c^n ...} = \cos (l\theta + m\phi + ...) - i \sin (l\theta + m\phi + ...).$$

$$\text{Hence } a^l b^m c^n ... + \frac{1}{a^l b^m c^n ...} = 2 \cos (l\theta + m\phi + ...).$$

4.3 POLAR FORM

Let $z = a + ib$. On putting $a = r \cos \theta$, $b = r \sin \theta$; we get

$z = r (\cos \theta + i \sin \theta)$,

$\Rightarrow r = \sqrt{a^2 + b^2}$ and $\tan \theta = b/a$.

Now $z = r(\cos\theta + i\sin\theta)$ and by De Moivre's Theorem,

$z^n = r^n(\cos n\theta + i\sin n\theta)$, $= r = \sqrt{a^2+b^2}$, $\theta = \tan^{-1}(b/a)$.

Example 1:

If n is a positive integer, show that

$$\left(\sqrt{3}+i\right)^n + \left(\sqrt{3}-i\right)^n = 2^{n+1}\cos(n\pi/6).$$

Solution :

Let $\sqrt{3} = r\cos\theta$ and $1 = r\sin\theta$ so that

$r^2 = 4$ and $\tan\theta = 1/\sqrt{3} \Rightarrow r = 2, \theta = \pi/6$.

$\therefore \left(\sqrt{3+i}\right)^n = 2^n(\cos\pi/6 + i\sin\pi/6)^n$

$\Rightarrow \left(\sqrt{3+i}\right)^n = 2^n\{\cos(n\pi/6) + i\sin(n\pi/6)\}$. ...(1)

Similarly $\left(\sqrt{3-i}\right)^n = 2^n\{\cos(n\pi/6) - i\sin(n\pi/6)\}$. ...(2)

Adding (1) and (2), we obtain

$\left(\sqrt{3+i}\right)^n + \left(\sqrt{3-i}\right)^n = 2.2^n\cos(n\pi/6) = 2^{n+1}\cos(n\pi/6)$.

Example 2:

If n be a positive integer, prove that

$$\left(1+i\sqrt{3}\right)^n + \left(1-i\sqrt{3}\right)^n = 2^{n-1}\cos\frac{n\pi}{3}.$$

Solution :

Let $r\cos\theta = 1$ and $r\sin\theta = \sqrt{3}$.

$\Rightarrow r^2 = 4$ and $\tan\theta = \sqrt{3} \Rightarrow r = 2, \theta = \pi/3$.

$\therefore 1 \pm i\sqrt{3} = 2\left(\cos\frac{\pi}{3} \pm i\sin\frac{\pi}{3}\right)$.

Now $\left(1+i\sqrt{3}\right)^n + \left(1-i\sqrt{3}\right)^n$

$= 2^n(\cos\pi/3 + i\sin\pi/3)^n + 2^n(\cos\pi/3 - i\sin\pi/3)^n$

$= 2^n\left(\cos\frac{n\pi}{3} + i\sin\frac{n\pi}{3}\right) + 2^n\left(\cos\frac{n\pi}{3} - i\sin\frac{n\pi}{3}\right)$

$$= 2^n \left(2 \cos \frac{n\pi}{3}\right) = 2^{n+1} \cos \frac{n\pi}{3}.$$

Example 3:

Prove that $(1 + i)^n + (1 - i)^n = 2^{\frac{n}{2}+1} \cos (n\pi/4)$.

Solution :

$1 = r \cos \theta$ and $1 = r \sin \theta \Rightarrow r^2 = 2$ and $\tan \theta = 1$.

$\Rightarrow r = \sqrt{2}$ and $\theta = \pi/4$, and so

$$1 + i = \sqrt{2} (\cos \pi/4 + i \sin \pi/4)$$

$\Rightarrow (1 + i)^n = 2^{n/2} (\cos \pi/4 + i \sin \pi/4)^n = 2^{n/2} [\cos n\pi/4 + i \sin n\pi/4]$ ...(1)

Similarly $(1 - i)^n = 2^{n/2} [\cos n\pi/4 - i \sin n\pi/4]$. ...(2)

Adding (1) and (2), we obtain

$$(1 + i)^n + (1 - i)^n = 2^{n/2} \cos (n\pi/4) = 2^{\frac{n}{2}+1} \cos (n\pi/4).$$

Example 4:

If α, β, be the rots of $x^2 - 2x + 2 = 0$, *prove that*

$$\alpha^n + \beta^n = 2^{\frac{n}{2}+1} \cos (n\pi/4).$$

Solution :

Solving $x^2 - 2x + 2 = 0$, we obtain

$$x = \frac{2 \pm \sqrt{4-8}}{2} = 1 + i.$$

Thus $a = 1 + i$ and $b = 1 - i$ and so

$$\alpha^n + \beta^n = (1 + i)^n + (1 - i)^n$$

$$= 2^{\frac{n}{2}+1} \cos (n\pi/4).$$

Example 5:

If $(1 + z)^n = c_0 + c_1 z + c_2 z^2 + \ldots + c_n z^n$ *where n is is a positive integer, show that*

$$c_0 - c_2 + c_4 - \ldots = 2^{n/2} \cos (n\pi/4),$$

$$c_1 - c_3 + c_5 - \ldots = 2^{n/2} \sin (n\pi/4).$$

Solution :

We have

$(1 + z)^n = c_0 + c_1z + c_2z^2 + c_3z^3 + c_4z^4 + ... + c_nz^n.$

Putting z = i, we obtain

$(1 + i)^n = c_0 + c_1i + c_2i^2 + c_3i^3 + c_4i^4 + c_5i^5 + ...$

$= c_0 + c_1i - c_2 - c_3i + c_4 + c_5i.....$

$\therefore (1 + i)^n = (c_0 - c_2 + c_4 - ....) + i\,(c_1 - c_2 + c_5 - .....)$. ...(1)

Let $1 + r \cos \theta$ and $1 = r \sin \theta$ so that

$r^2 = 2$ and $\tan \theta = 1 \Rightarrow \quad r = \sqrt{2}$ and $\theta = \pi/4$.

$\therefore 1 + i = r(\cos \theta + i \sin \theta) = \sqrt{2}\ (\cos \pi/4 + i \sin \pi/4)$

$\Rightarrow (1 + i)^n = 2^{n/2} (\cos \pi/4 + i \sin \pi/4)^n$

$$\Rightarrow (1 + i)^n = 2^{n/2}\left[\cos\frac{n\pi}{4} + i \sin\frac{n\pi}{4}\right]. \qquad ...(2)$$

Equating real and imaginary parts on the R.H.S. of (1) and (2), we obtain

$c_0 - c_2 + c_4 - ..... = 2^{n/2} \cos (n\pi/4),$

and $\quad c_1 - c_3 + c_5 - ..... = 2^{n/2} \sin (n\pi/4).$

Example 6:

If $(a_1 + ib_1)(a_2 + ib_2).....(a_n + ib_n) = A + iB$, prove that

(i) $\left(a_1^2 + b_1^2\right)\left(a_2^2 + b_2^2\right).....\left(a_n^2 + b_n^2\right) = A^2 + B^2$

(ii) $\tan^{-1}\frac{b_1}{a_1} + \tan^{-1}\frac{b_2}{a_2} + ...+ \tan^{-1}\frac{b_n}{a_n} = \tan^{-1}\frac{B}{A}.$

Solution :

Let $a_1 + ib_1 = r_1(\cos \theta_1 + i \sin \theta_1)$, so that

$a_1 = r_1 \cos b_1$ and $\theta_1 = r_1 \sin \theta_1$.

$\Rightarrow r_1^2 = a_1^2 + b_1^2$ and $\tan \theta_1 = b_1/a_1 \Rightarrow \theta_1 = \tan^{-1} b_1/a_1$.

Similarly $a_2 + ib_2 = r_2(\cos \theta_2 + i \sin \theta_2)$ gives

$r_2^2 = a_2^2 + b_2^2$ and $\theta_2 = \tan^{-1} b_2/a_2$ and so on.

Consider $(a_1 + ib_1)(a_2 + ib_2)....(a_n + ib_n) = A + iB$.

$\Rightarrow r_1(\cos \theta_1 + i \sin \theta_1).r_2(\cos \theta_2 + i \sin \theta_2)...r_n(\cos \theta_n + i \sin \theta_n)$

$= A + iB.$

$\Rightarrow r_1 r_2 \ldots . r_n (\cos \theta_1 + i \sin \theta_1)(\cos \theta_2 + i \sin \theta_2) + i \sin \theta_2)$
$\ldots(\cos \theta_n + i \sin \theta_n)$

$= A + iB.$

$\Rightarrow r_1 r_2 \ldots . r_n [\cos (\theta_1 + \theta_2 + \ldots . + \theta_n) + i \sin (\theta_1 + \theta_2 + \ldots + \theta_n)]$

$= A + iB.$

Comparing real and imaginary parts on both sides, we get

$r_1 r_2 \ldots . r_n \cos (\theta_1 + \theta_2 + \ldots . + \theta_n) = A,$ ...(1)

$r_1 r_2 \ldots . r_n \sin (\theta_1 + \theta_2 + \ldots . + \theta_n) = B,$ ...(2)

Squaring and adding (1) and (2), we get

$r_1^2 r_2^2 \ldots . r_n^2 = A^2 + B^2.$

Hence $\left(a_1^2 + b_1^2\right)\left(a_2^2 + b_2^2\right) \ldots \ldots \left(a_n^2 + b_n^2\right) = A^2 + B^2.$

Dividing (2) by (1), we get

$\tan (\theta_1 + \theta_2 + \ldots + \theta_n) = B/A$

$\Rightarrow \theta_1 + \theta_2 + \ldots . + \theta_n = \tan^{-1} (B/A).$

Hence $\tan^{-1} (b_1/a_1) + \tan^{-1} (b_2/a_2) + \ldots + \tan^{-1} (b_n/a_n)$

$= \tan^{-1} (B/A).$

Example 7:

If the roots of $t^2 - 2t + 2 = 0$ are α and β, s how that

(a) $$\frac{(x+\alpha)^n + (x+\beta)^n}{\alpha+\beta} = \frac{\cos n\theta}{\sin^n \phi}$$

(b) $$\frac{(x+\alpha)^n + (x+\beta)^n}{\alpha+\beta} = \frac{\cos n\theta}{\sin^n \phi}, \text{ where } x + 1 = \cot \phi.$$

Solution :

Solving $t^2 - 2t + 2$, we get

$$\Rightarrow t = \frac{2 \pm \sqrt{4-8}}{2} = \frac{2 \pm 2i}{2} = 1 \pm i.$$

Let $\alpha = 1 + i$ and $\beta = 1 - i$.

$\therefore x + \alpha = (x + 1) + i = r (\cos \phi + i \sin \phi),$

and $x + \beta = (x + 1) - i = r (\cos \phi - i \sin \phi),$

where $r \cos \phi = x + 1$, $r \sin \phi = 1$.

$\therefore \cot \phi = (x + 1)$ and $r = \sqrt{1+(x+1)^2} = \sqrt{1+\cot^2 \phi} = \text{coses } \phi.$

(a) $\dfrac{(x+\alpha)^n - (x+\beta)^n}{\alpha-\beta}$

$$= \frac{r^n(\cos n\phi + i \sin n\phi) - r^n(\cos n\phi - i \sin n\phi)}{\alpha-\beta}$$

$$= \frac{r^n(2i \sin n\phi)}{1+i-1+i} = r^n \sin n\phi$$

$$= \frac{\sin n\phi}{\sin^n \phi}. \quad (\because r \sin\phi = 1)$$

(b) $\dfrac{(x+\alpha)^n + (x+\beta)^n}{\alpha+\beta}$

$$= \frac{r^n(\cos n\phi + i \sin n\phi) + r^n(\cos n\phi - i \sin n\phi)}{1+i+1-i}$$

$$= r^n \cos nf = \frac{\cos n\phi}{\sin^n \phi}.$$

5

De Moivre's Theorem for a Fractional Index

5.1 FRACTIONAL INDEX

Theorem 1:

State and prove De Moivre's Theorem for a fractional index.

Or

If p and q be integers ($q \neq 0$), then cos (p^{θ}/q) + i sin (p^{θ}/q) is one of the values of (cos θ + i sin θ)$^{p/q}$.

Proof:

By De Moivre's Theorem for an integral index,

$$\left(\cos\frac{\theta}{q} + i\sin\frac{\theta}{q}\right)^q = \cos q.\frac{\theta}{q} + i\sin q.\frac{\theta}{q}$$

$$i = \cos\theta + i\sin\theta.$$

So the qth power of cos θ/q + i sin θ/q is cos θ + i sin θ,

i.e., (cos θ/q + i sin θ/q) is one of the qth roots of cos θ + i sinθ,

i.e., (cos θ/q + i sin θ/q) is one of the values of (cos θ + i sin θ)$^{1/q}$.

Raising both sides to the power p, we see that

(cos θ/q + i sin θ/q)^p is one of the values of (cos θ + i sin θ)$^{p/q}$.

Hence cos p^{θ}/q + i sin p^{θ}/q is one of the values of

(cos θ + i sin θ)$^{p/q}$. *(De Moivre's Theorem for an integral index)*

Theorem 2:

If p and q be integers prime to each other ($\theta \times 0$), then all the values of $(\cos\theta + i\sin\theta)^{p/q}$ are given by

$$\cos\left[(2k\pi + \theta)\frac{p}{q} + i\sin(2k\pi + \theta)\frac{p}{q}\right],$$

where k = 0, 1, 2,...., q − 1.

The proof is omitted.

Proof:

$$\left(\cos\frac{\pi}{3} + i\sin\frac{\pi}{3}\right)^{3/5}$$

$$= \cos\left(2k\pi + \frac{\pi}{3}\right).\frac{3}{5} + i\sin\left(2k\pi + \frac{\pi}{3}\right).\frac{3}{5},$$

where k = 0, 1, 2, 3, 4.

Example 1:

Write down all the values of $(i)^{1/6}$.

Solution :

We have $i = 0 + i.1 = \pi/2 + i\sin\pi/2$. Thus

$(i)^{1/6} = (\cos\pi/2 + i\sin\pi/2)^{1/6}$.

By Theorem 3,

$$(i)^{1/6} = \left[\cos\left(2k\pi + \frac{\pi}{2}\right)\frac{1}{6} + i\sin\left(2k\pi + \frac{\pi}{2}\right)\frac{1}{6}\right],$$

where k = 0, 1, 2, 3, 4, 5

$$= \cos(4k + 1)\frac{\pi}{12} + i\sin(4k + 1)\frac{\pi}{12}$$

$$= \cos\left(\frac{p\pi}{12}\right) + i\sin\left(\frac{p\pi}{12}\right),\ p = 1, 5, 9, 13, 17, 21.$$

Example 2:

Write down all the values of

(a) $\left(1 + i\sqrt{3}\right)^{2/3}$, *(b)* $\left(\sqrt{3} - i\right)^{3/5}$, *(c)* $(1 + i)^{1/3}$.

Solution :

(a) Let $1 + i\sqrt{3} = r(\cos\theta + i\sin\theta)$.

Then $1 = r\cos\theta$ and $\sqrt{3} = r\sin\theta$.

$\Rightarrow r^2 = 4$ and $\tan\theta = \sqrt{3} \Rightarrow r = 2, \theta = \pi/3$.

Now $\left(1+\sqrt{3}\right)^{2/3} = r^{2/3}(\cos\theta + i\sin\theta)^{2/3}$

$= 2^{2/3}(\cos\pi/3 + i\sin\pi/3)^{2/3}$

$$= 2^{2/3}\left[\cos\frac{2}{3}\left(2k\pi + \frac{\pi}{3}\right) + i\sin\frac{2}{3}\left(2k\pi + \frac{\pi}{3}\right)\right], \; k = 0, 1, 2.$$

$$= 2^{2/3}\left[\cos\frac{2(6k+1)\pi}{9} + i\sin\frac{2(6k+1)\pi}{9}\right], \; k = 0, 1, 2.$$

Hence $\left(1+i\sqrt{3}\right)^{2/3} = 2^{2/3}\left[\cos\frac{p\pi}{9} + i\sin\frac{p\pi}{9}\right]$, $p = 2, 14, 26$.

(b) Let $\sqrt{3}-1 = r(\cos\theta - i\sin\theta)$.

Then $\sqrt{3} = r\cos\theta$, $1 = r\sin\theta$

$\Rightarrow r^2 = 4$ and $\tan\theta = 1/\sqrt{3} \Rightarrow r = 2$ and $\theta = \pi/6$.

Now $\left(\sqrt{3}-i\right)^{3/5} = (2)^{3/5}(\cos\pi/6 - i\sin\pi/6)^{3/5}$

$= (2)^{3/5}(\cos\pi/2 - i\sin\pi/2)^{1/5}$. *(De Moivre's Theorem)*

$$= (2)^{3/5}\left[\cos\frac{3\pi}{2} + i\sin\frac{3\pi}{2}\right]^{1/5}$$

$$= (2)^{3/5}\left[\cos\left(2k\pi + \frac{3\pi}{2}\right)\frac{1}{5} + i\sin\left(2k\pi + \frac{3\pi}{2}\right)\frac{1}{5}\right],$$

$k = 0, 1, 2, 3, 4$

$$= (2)^{3/5}\left[\cos(4k+3)\frac{\pi}{10} + i\sin(4k+3)\frac{\pi}{10}\right], \; k = 0, 1, 2, 3, 4$$

Hence $\left(\sqrt{3}-i\right)^{3/5} = (2)^{3/5}\cos\frac{p\pi}{10} + i\sin\frac{p\pi}{10}$, $p = 3, 7, 11, 15, 19$.

(c) Let $1 + i = r(\cos\theta + i\sin\theta)$.

Then $1 = r\cos\theta$ and $1 = r\sin\theta$.

$\Rightarrow$ $r^2 = 2$ and $\tan\theta = 1 \Rightarrow r = \sqrt{2}$ and $\theta = \pi/4$.

Now $(1 + i)^{1/3} = \left(\sqrt{2}\right)^{1/3}(\cos\pi/4 + i\sin\pi/4)^{1/3}$.

$$= \left(2^{1/2}\right)^{1/3}\left[\cos\frac{1}{3}\left(2k\pi+\frac{\pi}{4}\right)+i\sin\frac{1}{3}\left(2k\pi+\frac{\pi}{4}\right)\right], k = 0, 1, 2,$$

$$= 2^{1/6}\left[\cos(8k+1)\frac{\pi}{12}+i\sin(8k+1)\frac{\pi}{12}\right], k = 0, 1, 2.$$

Hence $(1 + i)^{1/3} = 2^{1/6}\left[\cos\left(\frac{p\pi}{12}\right)+i\sin\left(\frac{p\pi}{12}\right)\right], \pi = 1, 9, 17.$

Example 3:

Find all the values of $\left(\frac{1}{2}+i\frac{\sqrt{3}}{2}\right)^{3/4}$ *and show that the continued product of all the values is 1.*

Solution :

Let $\left(\frac{1}{2}+i\frac{\sqrt{3}i}{2}\right) = r(\cos\theta + i\sin\theta).$

$\therefore 1/2 = r\cos\theta$ and $\sqrt{3/2} = r\sin\theta,$

$\Rightarrow r^2 = 1$ and $\tan\theta = \sqrt{3} \Rightarrow r = 1$ and $\theta = \pi/3.$

Now $\left(\frac{1}{2}+\frac{\sqrt{3}}{2}\right)^{3/4} = \left[\left(\cos\pi/3+i\sin\pi/3\right)^3\right]^{1/4}$

$= (\cos\pi + i\sin\pi)^{1/4}$

$= [\cos(2k\pi + \pi) + i\sin(2k\pi + \pi)^{1/4}, k = 0, 1, 2, 3$

$= \cos(2k+1)\frac{\pi}{4} + i\sin(2k+1)\frac{\pi}{4}, k = 0, 1, 2, 3.$

Hence the required four values are, respectively

$\cos(\pi/4) + i\sin(\pi/4)$, $\cos(3\pi/4) + i\sin(3\pi/4)$,

$\cos(5\pi/4) + i\sin(5\pi/4)$ and $\cos(7\pi/4) + i\sin(7\pi/4)$.

The combined product of the above four values

$$= \cos\left(\frac{\pi}{4}+\frac{3\pi}{4}+\frac{7\pi}{4}\right)+\sin\left(\frac{\pi}{4}+\frac{3\pi}{4}+\frac{5\pi}{4}+\frac{7\pi}{4}\right)$$

$= \cos 4\pi + i\sin 4\pi = 1 + i . 0 = 1.$

Example 4:

Prove that n nth roots of unity form a series in G.P.

Solution :

Let $x = (1)^{1/n} = (\cos 0 + i \sin 0)^{1/n}$

$$= \cos\left(\frac{0 + 2k\pi}{n}\right) + i \sin\left(\frac{0 + 2k\pi}{n}\right) \quad k = 0, 1, 2, \ldots n - 1$$

Thus n roots of unity are

$$1, \cos\left(\frac{2\pi}{n}\right) + i \sin\left(\frac{2\pi}{n}\right), \cos\left(\frac{4\pi}{n}\right) + i \sin\left(\frac{4\pi}{n}\right), \ldots\ldots,$$

$$\cos\left\{\frac{2}{n}(n-1)\pi\right\} + i \sin\left\{\frac{2}{n}(n-1)\pi\right\}$$

$\Rightarrow 1, \alpha, \alpha^2, \ldots, \alpha^{n-3}$; where $\alpha = \cos\left(\frac{2\pi}{n}\right) + 1 \sin\left(\frac{2\pi}{n}\right)$

The above roots $1, \alpha, \alpha^2, \ldots \alpha^{n-1}$ are clearly G.P.

Example 5:

Find the fifth roots of (–32).

Solution :

We have $(-32)^{1/5} = \{2^5(-1)\}^{1/5} = 2(-1)^{1/5}$

$$= 2(\cos \pi + i \sin \pi)^{1/5}$$

$$= 2\left[\cos \frac{1}{5}(2k\pi + \pi) + i \sin \frac{1}{5}(2k\pi + \pi)\right], k = 0, 1, 2, 3, 4.$$

Hence $(-32)^{1/5} = 2\{\cos(p\pi/5) + i \sin(p\pi/5)\}$, $p = 1, 3, 5, 7, 9$.

Example 6:

We have $(16)^{1/4} = (2^4)^{1/4} = 2(1)^{1/4}$

$= 2(\cos 0 + i \sin 0)^{1/4}$

$$= 2\left\{\cos \frac{1}{4}(2k\pi + 0) + i \sin \frac{1}{4}(2k\pi + 0)\right\}, k = 0, 1, 2, 3$$

$= 2 \times 1, 2 \times i, 2 \times -1, 2 \times -i = \pm 2, \pm 2i.$

Example 7:

Show that one of the values of

$$(a + ib)^{m/n} + (a - ib)^{m/n} \text{ is } 2(a^2 + b^2)^{m/2n} \cos\left(\frac{m}{n} \tan^{-1} \frac{b}{a}\right).$$

Solution :

Let $a + ib = r(\cos\theta + i\sin\theta)$. ...(i)

Then $r\cos\theta = a$ and $r\sin\theta = b$.

$\Rightarrow$ $r^2 = a^2 + b^2$, $\tan\theta = b/a$

$\Rightarrow$ $\theta = ta^{-1}\ b/a$. ...(ii)

Taking conjugates of both sides of (i), we get

$a - ib = r(\cos\theta - i\sin\theta)$.

Now by De Moivre's Theorem, one of the values of

$$(a + ib)^{m/n} \text{ is } r^{m/n}\left(\cos\frac{m\theta}{n} + i\sin\frac{m\theta}{n}\right),$$

and one of the values of

$$(a - ib)^{m/n} \text{ is } r^{m/n}\left(\cos\frac{m\theta}{n} - i\sin\frac{m\theta}{n}\right).$$

Hence one of the values of $(a + ib)^{m/n} + (a - ib)^{m/n}$ is

$$= 2r^{m/n}\cos\frac{m\theta}{n}$$

$$= 2(a^2 + b^2)^{m/2n}\cos\left(\frac{m}{n}\tan^{-1}\frac{b}{a}\right), \text{ using (ii).}$$

Example 8:

Write down all the values of each of the following :

(i) $i^{1/3}$ *(ii)* $(1 + i)^{1/4}$ *(iii)* $\left(\sqrt{3+i}\right)^{1/3}$.

Ans.

(i) $\cos(\pi/12) + i\sin(\pi/12)$, $\pi = 1, 5, 9, 13, 17, 21$.

(ii) $2^{1/8}\{\cos(p\pi/16) + i\sin(p\pi/16)\}$, $p = 1, 9, 17, 25$.

(ii) $2^{1/3}\{\cos(p\pi/18) + i\sin(p\pi/18)\}$, $\pi = 1, 13, 25$.

5.2 SOLUTION OF EQUATIONS

In this section we shall solve certain equations with the help of De Moivre's Theorem.

Example 1:

Solves the equations :

(i) $x^3 - 1 = 0$ *(ii)* $x^3 + 1 = 0$.

Solution :

(i) $x^3 - 1 = 0 \Rightarrow x^3 = 1 = \cos 0 + i \sin 0$

$\Rightarrow x = \{\cos (2k\pi + 0) + i \sin (2k\pi + 0)\}^{1/8}$

$\Rightarrow x = \cos \frac{2}{3} k\pi + i \sin \frac{2}{3} k\pi$, k = 0, 1, 2.

Hence $1, -\frac{1}{2} + i \frac{\sqrt{3}}{2}, -\frac{1}{2} - i \frac{\sqrt{3}}{2}$ are the roots.

(ii) $x^3 + 1 = 0 \Rightarrow x^2 = -1 \Rightarrow x = (\cos \pi + i \sin \pi)^{1/3}$

$\Rightarrow x = \cos \frac{1}{2} (2k\pi + \pi) + i \sin \frac{1}{2} (2k\pi + \pi)$, k = 0, 1, 2.

Hence $\frac{1}{2} + i \frac{\sqrt{3}}{2}, -1, \frac{1}{2} - i \frac{\sqrt{3}}{2}$ are the roots.

Example 2:

Solve the equation $z^4 + 1 = 0$.

Solution :

$z^4 + 1 = 0 \Rightarrow = -1$

$\Rightarrow z = (\cos \pi + i \sin \pi)^{1/4}$

$\Rightarrow z = \cos \frac{1}{4} (2k\pi + \pi) + i \sin \frac{1}{4} (2k\pi + \pi)$, k = 0, 1, 2, 3.

$\Rightarrow z = \cos \pi/4 + i \sin \pi/4, \cos (3\pi/4) + i \sin (3\pi/4)$,

$\cos (5\pi/4) + i \sin (5\pi/4), \cos (7\pi/4) + i \sin (7\pi/4)$

$\frac{1}{\sqrt{2}} (1 + i), \frac{1}{\sqrt{2}} (-1 + i), \frac{1}{\sqrt{2}} (-1 - i), \frac{1}{\sqrt{2}} (1 - i)$.

Hence the four roots of $z^4 + 1 = 0$ are $\frac{1}{\sqrt{2}} (\pm 1 \pm i)$.

Example 3:

Solve the equation $z^5 + 1 = 0$.

Solution :

We have $z^5 + 1 = 0$

$\Rightarrow z^5 = -1 = \cos \pi + i \sin \pi$.

Therefore $z = (\cos \pi + i \sin \pi)^{1/5}$

$= \cos \frac{1}{5} (2k + 1) \pi + i \sin \frac{1}{5} (2k + 1) \pi$, k = 0, 1, 2, 3, 4.

Hence $\cos(\pi/5) + i\sin(\pi/5)$, $\cos(3\pi/5) + i\sin(3\pi/5)$, -1,
$\cos(7\pi/5) + i\sin(7\pi/5)$, $\cos(9\pi/5) + i\sin(9\pi/5)$
are the required roots.

Example 4:

Solve the equation $z^5 - 1 = 0$.

Solution :

$z^5 - 1 = 0 \Rightarrow z^5 = 1 = \cos 0 + i\sin 0$

$\Rightarrow z = (\cos 0 + i\sin 0)^{1/5} = (\cos 2k\pi + i\sin 2k\pi)^{1/5}$

Hence $z = \cos\left(\frac{2}{5}k\pi\right) + i\sin\left(\frac{2}{5}k\pi\right)$, $k = 0, 1, 2, 3, 4.$

Example 5:

Solve the equation $x^6 + 1 = 0$.

Solution :

We have $x^6 = -1 = \cos\pi + i\sin\pi$

$\Rightarrow x = [\cos(2k\pi + \pi) + i\sin(2k\pi + 1)^{1/6}$

$\Rightarrow x = [\cos(2k + \pi)\frac{\pi}{6} + i\sin(2k + 1)\frac{\pi}{6}$, $k = 0, 1, 2, 3, 4, 5.$

$= \cos\frac{p\pi}{6} + i\sin\frac{p\pi}{6}$, $p = 1, 3, 5, 7, 9, 11$

$= \cos\pi/6 + i\sin\pi/6, i, -\cos\pi/6 + i\sin\pi/6,$

$-\{\cos\pi/6 + i\sin\pi/6\}, -i, \cos\pi/6 - i\sin\pi/6.$

$[\because 5\pi/6 = \pi - \pi/6, 7\pi/6 = \pi + \pi/6]$

Hence the six values are

$\pm i, \frac{\sqrt{3}}{2} \pm \frac{i}{2}, -\frac{\sqrt{3}}{2} \pm \frac{i}{2}.$

Example 6:

Solve the equation

(i) $z^7 - z = 0$ (ii) $z^5 - z = 0$.

Solution :

(i) $z^7 - z = 0 \Rightarrow z(z^6 - 1) = 0$

$\Rightarrow z = 0 \Rightarrow z^6 = 1.$

Now $z^6 = 1 = \cos 0 + i\sin 0$

$$\therefore z = \cos\left(\frac{0+2k\pi}{6}\right) + i\sin\left(\frac{0+2k\pi}{6}\right),\ k = 0, 1, 2, 3, 4, 5.$$

$$= \cos\left(\frac{k\pi}{3}\right) + i\sin\left(\frac{k\pi}{3}\right),\ k = 0, 1, 2, 3, 4, 5.$$

Hence $1, \frac{1}{2} + i\frac{\sqrt{3}}{2}, -\frac{1}{2} + i\frac{\sqrt{3}}{2}, -1, -\frac{1}{2} - i\frac{\sqrt{3}}{2}$

$\frac{1}{2} - i\frac{\sqrt{3}}{2}$ and 0 are the required seven roots of $z^7 - z = 0$.

(ii) $z^5 - z = z(z^4 - 1) = z(z^2 - 1)(z^2 + 1)$.

Hence the roots of $z^5 - z = 0$ are $0, \pm 1, \pm i$.

Example 7:

Solve the equation $x^7 - 1 = 0$.

Solution :

$x^7 - 1 = 0 \Rightarrow x = (1)^{1/7} = (\cos 0 + i \sin 0)^{1/7}$

$$\therefore x = \cos\frac{2k\pi}{7} + i\sin\frac{2k\pi}{7},\ k = 0, 1, 2, 3, 4, 5, 6.$$

Thus the seven values are

$\cos 0 + i\sin 0$, $\cos(2\pi/7) + i\sin(2\pi/7)$, $\cos(4\pi/7) + i\sin(4\pi/7)$,

$\cos(6\pi/7) + i\sin(6\pi/7)$, $\cos(8\pi/7) + i\sin(8\pi/7)$,

$\cos(10\pi/7) + i\sin(10\pi/7)$, $\cos(12\pi/7) + i\sin(12\pi/7)$.

Now $\cos\frac{8\pi}{7} + i\sin\frac{8\pi}{7} = \cos\left(2\pi - \frac{6\pi}{7}\right) + i\sin\left(2\pi - \frac{6\pi}{7}\right)$

$$= \cos\frac{6\pi}{7} - i\sin\frac{6\pi}{7}.$$

Similarly $\cos\frac{10\pi}{7} + i\sin\frac{10\pi}{7} = \cos\frac{4\pi}{7} - i\sin\frac{4\pi}{7}$.

and $\cos\frac{12\pi}{7} + i\sin\frac{12\pi}{7} = \cos\frac{2\pi}{7} - i\sin\frac{2\pi}{7}$.

Hence the seven values are

$$1, \left(\cos\frac{2\pi}{7} \pm i\sin\frac{2\pi}{7}\right), \left(\cos\frac{4\pi}{7} \pm i\sin\frac{4\pi}{7}\right)$$

and $\left(\cos\frac{6\pi}{7} \pm i\sin\frac{6\pi}{7}\right)$.

Example 8:

Solve the equations :

(a) $x^7 - x^4 + x^3 - 1 = 0.$

(b) $x^6 - x^5 + x^4 - x^3 + x^2 - x + 1 = 0.$

Solution :

(a) We see that

$$x^7 - x^4 + x^3 - 1 = 0 \Rightarrow x^4 (x^3 - 1) + (x^3 - 1) = 0$$

$$\Rightarrow (x^4 + 1)(x^3 - 1) = 0 \Rightarrow x^3 - 1 = 0 \Rightarrow x^4 + 1 = 0.$$

$$1, \frac{1}{2}\left(-1 \pm i\sqrt{3}\right), \frac{1}{\sqrt{2}} (\pm 1 \pm i).$$

(b) We have $x^6 - x^5 + x^4 - x^3 + x^2 - x + 1 = 0.$

Multiplying the given equation by $x + 1$, we get

$$x^7 + 1 = 0 \Rightarrow x^7 = -1 = \cos\pi + i\sin\pi.$$

$$\Rightarrow x = [\cos(2k\pi + \pi) + i\sin(2k\pi + \pi)]^{1/7}$$

$$= \cos(2k\pi + 1)\frac{\pi}{7} + i\sin(2k + 1)\frac{\pi}{7}, \ k = 0, 1, \ldots\ldots, 6$$

$$= \cos\left(\frac{p\pi}{7}\right) + i\sin\left(\frac{p\pi}{7}\right), \ p = 1, 3, 5, 7, 9, 11, 13$$

For $p = 7$, we get $x = -1$, which is a root of $x + 1 = 0$.

Hence the six values of x are

$$\cos\left(\frac{p\pi}{7}\right) + i\sin\left(\frac{p\pi}{7}\right), \ p = 1, 3, 5, 9, 11, 13.$$

Example 9:

Solve $x^6 + x^5 + x^4 + x^3 + x^2 + x + 1 = 0.$

Solution :

Multiplying the given equation by $(x - 1)$, we get

$$(x - 1)(x^6 + x^5 + x^4 + x^3 + x^2 + x + 1) = 0 \Rightarrow x^7 - 1 = 0.$$

Hence, by the required roots are

$$\cos\left(\frac{k\pi}{7}\right) \pm i\sin\left(\frac{k\pi}{7}\right), k = 2, 4, 6.$$

(Neglect the value $x = 1$ which corresponds to the factor $x - 1$).

Example 10:

Solve $x^9 + x^5 + x^4 + 1 = 0.$

Solution :

The given equation is $(x^5 + 1)(x^4 + 1) = 0$.

By the roots of $x^5 + 1 = 0$ are

$$\cos\left\{(2k+1)\frac{\pi}{5}\right\} + i\sin\left\{(2k+1)\frac{\pi}{5}\right\},\ k = 0, 1, 2, 3, 4. \qquad ...(A)$$

By the roots of $x^4 + 1 = 0$ are

$$\frac{1}{\sqrt{2}}(\pm 1 \pm i) \qquad ...(B).$$

Hence the values given in (A) and (B) are the required 9 roots of the given equation.

Example 11:

Solve $z^4 - z^3 + z^2 - z + 1 = 0.$

Solution :

Multiplying the given equation by $(z + 1)$ and simplying, we obtain, $2^5 + 1 = 0$, whose roots are

$z = \cos(p\pi/5) + \sin(p\pi/5)$, $p = 1, 3, 5, 7, 9,$

For $p = 5$, the root $z = -1$ corresponds to $z + 1 = 0$.

Hence the required roots are

$\cos(p\pi/5) + i\sin(p\pi/5)$; $p = 1, 3, 7, 9.$

Example 12:

Solve $z^5 + z^4 + z^3 + z^2 + z + 1 = 0.$

Solution :

$z^5 + z^4 + z^3 + z^2 + z + 1 = (z^4 - 1)/(z - 1).$

By the roots of $z^6 - 1 = 0$ are

$\cos(k\pi/3) + i\sin(k\pi/3)$; $k = 0, 1, 2, 3, 4, 5.$

Rejecting the root $z = 1$ (which corresponds to $k = 0$), the required roots are

$\cos(k\pi/3) + i\sin(k\pi/3)$; $k = 1, 2, 3, 4, 5.$

$$\Rightarrow -1, \pm\frac{1}{2} \pm i\frac{\sqrt{3}}{2}.$$

Example 13:

Solve the equation $z^8 - z^5 + z^3 - 1 = 0$.

Solution :

$z^8 - z^5 + z^3 - 1 = (z^5 + 1)(z^3 - 1)$.

Solving $z^3 - 1 = 0$ and $z^5 + 1 = 0$, the required roots are

$-1, -\frac{1}{2}\left(1 \pm i\sqrt{3}\right)$; $\cos(k\pi/5) + i\sin(k\pi/5)$, k = 1, 3, 5, 7, 9.

Example 14:

Solve the equation $z^{10} - 2^5 + 1 = 0$.

Solution :

Solving $z^{10} - z^5 + 1 = 0$ as a quadratic in z^5, we get

$$z^5 = \frac{1 \pm \sqrt{1-4}}{2} = \frac{1}{2} \pm i\frac{\sqrt{3}}{2} = \cos\frac{\pi}{3} \pm i\sin\frac{\pi}{3}$$

$\Rightarrow z = [\cos(2k\pi + \pi/3) \pm i\sin(2k\pi + \pi/3)]^{1/5}$.

Hence $z = \cos(6k+1)\frac{\pi}{15} \pm i\sin(6k+1)\frac{\pi}{15}$, k = 0, 1, 2, 3, 4 are the required roots.

Example 15:

Solve the equation $x^6 + x^3 + 1 = 0$.

Solution :

Solving $x^6 + x^3 + 1 = 0$ as a quadratic in x^3, we get

$$x^3 = \frac{-1 \pm \sqrt{1-4}}{2} = -\frac{1}{2} \pm i\frac{\sqrt{3}}{\sqrt{2}} = r(\cos\theta \pm i\sin\theta).$$

$\therefore r\cos\theta = -\frac{1}{2}$ and $r\sin\theta = \sqrt{3/2}$

$\Rightarrow r^2 = 1$ and $\tan\theta = -\sqrt{3}$.

$\Rightarrow r = 1$ and $\theta = \frac{2\pi}{3} \Rightarrow x^3 = \cos\frac{2\pi}{3} \pm i\sin\frac{2\pi}{3}$.

$$\therefore x = \left[\cos\left(2kp + \frac{2\pi}{3}\right) \pm i\sin\left(2k + \frac{2\pi}{3}\right)\right]^{1/2}$$

$$= \cos(6k+2)\frac{\pi}{9} \pm i\sin(6k+2)\frac{\pi}{9}, \; k = 0, 1, 2$$

Hence $\cos\left(\frac{p\pi}{9}\right) \pm i \sin\left(\frac{p\pi}{9}\right)$, p – 2, 8, 14

are the required roots.

Example 16:

Solve the equation $z^4 + 4z^2 + 16 = 0$.

Solution :

Solving $z^4 + 4z^2 + 16 = 0$ as a quadratic in z^2,

$$z^2 = \frac{-4 \pm \sqrt{16-64}}{2} = -2 \pm 2\sqrt{3}i$$

Let $z^2 = -2 \pm 2\sqrt{3}i = r(\cos\theta \pm i \sin\theta)$. ...(1)

$\therefore$ $r\cos\theta = -2$ and $r\sin\theta = 2\sqrt{3}$.

$\Rightarrow r^2 = 16$ and $\tan\theta = -\sqrt{3} \Rightarrow r = 4, \theta = 2\pi/3$. ...(2)

$$z = 2\left(\cos\frac{2\pi}{3} \pm i \sin\frac{2\pi}{3}\right)^{1/2}, \text{ by (1) and (2)}$$

$$= \left[\cos\left(2k\pi + \frac{2\pi}{3}\right).\frac{1}{2} \pm i \sin\left(2k\pi + \frac{2\pi}{3}.\frac{1}{2}\right)\right]; k = 0, 1.$$

$$= 2\left(\cos\frac{\pi}{3} \pm i \sin\frac{\pi}{3}\right), 2\left(\cos\frac{4\pi}{3} \pm i \sin\frac{4\pi}{3}\right),$$

$$= \left(\frac{1}{2} \pm i\frac{\sqrt{3}}{2}\right), -2\left(\frac{1}{2} \pm i\frac{\sqrt{3}}{2}\right)$$

Hence $\pm 1 \pm i\sqrt{3}$ are the required roots.

Example 17:

Find the seven seventh roots of unity and prove that for any positive integer n the sum of their nth powers vanishes, unless n is a multiple of 7 in which case the sum is 7.

Solution :

The seven seventh roots of unity *i.e.*,

$$x = (1)^{1/7} = (\cos 2k\pi + i \sin 2k\pi)^{1/7}$$

are $\alpha^k \equiv \cos(2k\pi/7) + i \sin(2k\pi/7)$, k = 0, 1, 2, ..., 6.

We have $\alpha = \cos(2\pi/7) + i \sin(2\pi/7)$.

The sum of the nth powers of these roots is

$$S = 1^n + \alpha^n + (\alpha^2)^n + (\alpha^3)^n + (\alpha^4)^n + (\alpha^5)^n + (\alpha^6)^n$$

$$\Rightarrow S = 1 + \alpha^n + \alpha^{2n} + \alpha^{3n} + \alpha^{4n} + \alpha^{5n} + \alpha^{6n}. \quad ...(1)$$

Case I: Let n be a mutliple of 7 so that n = 7m, m being a positive integer.

$$\therefore \alpha^{kn} = (\alpha^k)^{7m} = \{\cos(2k\pi/7) + i\sin(2k\pi/7)\}^{7m}$$

$$= \cos 2km\pi + i\sin 2km\pi$$

$$= (\cos 2m\pi + i\sin 2m\pi)^k = 1^k.$$

Thus $\alpha^{kn} = 1$ for k = 0, 1, 2, ..., 6.

Using in (1), S = 7.

Case II: Let n be not a multiple of 7 so that $\alpha^n \neq 1$.

From (1), $S = \dfrac{1-(\alpha^n)^7}{1-\alpha^n} = \dfrac{1-(\alpha^7)^n}{1-\alpha^n} = \dfrac{1-1}{1-a^n} \quad (\because \alpha^7 = 1)$

Hence $S = 0$, as $1 - a^n \neq 0$.

Example 18:

Show that the roots of the equation $(z-1)^5 + z^5 = 0$ are given by

$$z = \frac{1}{2}\left[1 + i\cos\left(\frac{p\pi}{10}\right)\right], p = 1, 3, 5, 7, 9.$$

Solution :

The given equation may be written as

$$\left(\frac{z-1}{z}\right)^5 + 1 = 0$$

$\Rightarrow w^5 + 1 = 0$, where

$$w = \frac{z-1}{z}$$

$$\Rightarrow z = \frac{1}{1-w}. \quad ...(1)$$

Now $w^5 + 1 = 0$

$\Rightarrow w^5 = -1 = \cos\pi + i\sin\pi.$

$\Rightarrow w = (\cos\pi + i\sin\pi)^{1/5}.$

$$\Rightarrow w = \cos\left\{\frac{2k\pi+\pi}{5}\right\} + i\sin\left\{\frac{2kp+\pi}{5}\right\}, k = 0, 1, 2, 3, 4.$$

We write $w = \cos\theta + i\sin\theta,\ \theta = (2k+1)\dfrac{\pi}{5}$.

Putting this value of of w in (1), we obtain

$$z = \frac{1}{1-\cos\theta - i\sin\theta}$$

$$= \frac{1}{1-\left(1-2\sin^2\theta/2\right)-2i\sin\theta/2\cos\theta/2}$$

$$= \frac{1}{-2i\sin\theta/2\left\{\cos^2\theta/2+i\sin\theta/2\right\}}$$

$$= \frac{\left(\cos\theta/2 - i\sin\theta/2\right)}{-2i\sin\theta/2} = \frac{1}{2} - \frac{1}{2i}\cot\frac{\theta}{2}$$

$$= \frac{1}{2}(1+i\cot q/2) \qquad (\because\ -1/i = i^2/i = i)$$

$$= \frac{1}{2}\left[1+i\cot\left\{\frac{(2k+1)\pi}{10}\right\}\right],\ k = 0, 1, 2, 3, 4.$$

$$= \frac{1}{2}\left[1+i\cot\left(\frac{p\pi}{10}\right)\right],\ p = 1, 3, 5, 7, 9.$$

Example 19:

Show that the roots of the equation

$$(1-z)^7 + (1+z)^7 = 0 \text{ are } z = \pm i\cot\left(\frac{p\pi}{7}\right),\ p - 1, 2, 3.$$

Solution :

We have

$$\left(\frac{1+z}{1-z}\right)^7 + 1 = 0 \Rightarrow w^7 + 1 = 0, \text{ where}$$

$$w = \frac{1+z}{1-z} \Rightarrow z = \frac{w-1}{w-1}. \qquad \text{...(1)}$$

Now $w^7 + 1 = 0 \Rightarrow w^7 = -1 = \cos\pi + i\sin\pi$

$\Rightarrow w = (\cos\pi + i\sin\pi)^{1/7}$.

$$\therefore w = \cos\left\{(2k+1)\frac{\pi}{7}\right\} + i\sin\left\{(2k+1)\frac{\pi}{7}\right\},\ k = 0, 1, 2, 3, 4, 5, 6.$$

$\Rightarrow w = \cos\theta + i\sin\theta$, where $\theta = (2k+1)\dfrac{\pi}{7}$.

Putting this value of w in (1), we obtain

$$z = \frac{\cos\theta + i\sin\theta - 1}{1 + \cos\theta + i\sin\theta}$$

$$= \frac{1 - 2\sin^2\theta/2 + 2i\sin\theta/2\cos\theta/2 - 1}{1 + 2\cos^2\theta/2 - 1 + 2i\sin\theta/2\cos\theta/2}$$

$$= \frac{2i\sin\theta/2\{\cos\theta/2 + i\sin\theta/2\}}{2\cos\theta/2\{\cos\theta/2 + i\sin\theta/2\}}$$

$$= i\tan\frac{\theta}{2} = i\sin\left\{(2k+1)\frac{\pi}{14}\right\},\ k = 0, 1, 2, 3, 4, 5, 6.$$

For k = 3, we get an infinite value of z which is rejected.

Thus $z = i\tan\left(\dfrac{p\pi}{14}\right)$, where p = 1, 3, 5, 9, 11, 13. ...(2)

We observe that

$$\tan\left(\frac{\pi}{14}\right) = \tan\left(\frac{\pi}{2} - \frac{3\pi}{7}\right) = \cot\frac{2\pi}{7};$$

$$\tan\left(\frac{3\pi}{14}\right) = \tan\left(\frac{\pi}{2} - \frac{2\pi}{7}\right) = \cot\frac{2\pi}{7};$$

$$\tan\left(\frac{5\pi}{14}\right) = \tan\left(\frac{\pi}{2} - \frac{\pi}{7}\right) = \cot\frac{\pi}{7}$$

$$\tan\left(\frac{9\pi}{14}\right) = \tan\left(\frac{\pi}{2} + \frac{\pi}{7}\right) = -\cot\frac{\pi}{7};$$

$$\tan\left(\frac{11\pi}{14}\right) = -\cot\frac{2\pi}{7},\ \tan\left(\frac{13\pi}{4}\right) = -\cot\frac{3\pi}{7}.$$

Hence, by (2), the required roots are

$$z = \pm i\cot\left(\frac{p\pi}{7}\right),\ p = 1, 2, 3.$$

Example 20:

Solve the equation

$(z+1)^7 = z^7(\cos 7\theta + i\sin 7\theta)$.

Solution :

The given equation is $w^7 = \cos 7\theta + i \sin 7\theta$; ...(1)

$$w = \frac{z+1}{z} \Rightarrow z = \frac{1}{w-1}. \quad \text{From (1),}$$

$$w = \cos\left(\frac{2k\pi + 7\theta}{7}\right) + i \sin\left(\frac{2k\pi + 7\theta}{7}\right), k = 0, 1, ..., 6$$

$\Rightarrow w = \cos\alpha + i \sin\alpha$, $\alpha = \theta + (2/7)\, k\pi$.

$$\therefore z = \frac{1}{\cos\alpha + i\sin\alpha - 1}$$

$$= \frac{1}{\{1 - 2\sin^2(\alpha/2)\} + 2i\sin(\alpha/2)\cos(\alpha/2) - 1}$$

$$= \frac{1}{2i\sin(\alpha/2)\{\cos(\alpha/2) + i\sin(\alpha/2)\}}$$

$$= \frac{\cos(\alpha/2) - i\sin(\alpha/2)}{2i\sin(\alpha/2)} = -\frac{1}{2} = \frac{1}{2i}\cot\left(\frac{\alpha}{2}\right)$$

$$= -\frac{1}{2} + \frac{1}{2i^2}\cot\left(\frac{\alpha}{2}\right) \quad (i^2 = -1)$$

$$= -\frac{1}{2}\left\{1 + i\cot\left(\frac{\alpha}{2}\right)\right\}$$

$$= -\frac{1}{2}\left\{1 + i\cot\left(\frac{\alpha}{2} + k\frac{\pi}{7}\right)\right\}, k = 0, 1....., 6.$$

Example 21:

Solve the equation $(2 + z)^6 + (2 - z)^6 = 0$.

Solution :

The given equation can be written as

$$w^6 + 1 = 0, \text{ where } w = \frac{2+z}{2-z} \quad ...(1)$$

$\Rightarrow w^6 = -1 = \cos\pi + i\sin\pi$.

$$\therefore w = \left\{\cos\left(\frac{2k\pi + \pi}{6}\right) + i\sin\left(\frac{2k\pi + \pi}{6}\right)\right\}, k = 0, 1, ..., 5$$

$\Rightarrow w = \cos\theta + i\sin\theta$, where $\theta = (2k+1)\dfrac{\pi}{6}$.

From (1), $(2 - z)\,w = 2 + z \Rightarrow z = \dfrac{2(w-1)}{w+1}$

$$\Rightarrow z = \frac{2(\cos\theta + i\sin\theta - 1)}{(\cos\theta + i\sin\theta + 1)}$$

$$= \frac{\cos(\alpha/2) - i\sin(\alpha/2)}{2i\sin(\alpha/2)} = -\frac{1}{2} = \frac{1}{2i}\cot\left(\frac{\alpha}{2}\right)$$

$$= -\frac{1}{2} + \frac{1}{2i^2}\cot\left(\frac{\alpha}{2}\right) \quad (i^2 = -1)$$

$$= -\frac{1}{2}\left\{1 + i\cot\left(\frac{\alpha}{2}\right)\right\}$$

$$= -\frac{1}{2}\left\{1 + i\cot\left(\frac{\alpha}{2} + k\frac{\pi}{7}\right)\right\}, \; k = 0, 1 \ldots\ldots, 6.$$

Example 22:

Solve the equation $(2 + z)^6 + (2 - z)^6 = 0$.

Solution :

The given equation can be written as

$$w^6 + 1 = 0, \text{ where } w = \frac{2+z}{2-z} \qquad \ldots(1)$$

$\Rightarrow w^6 = -1 = \cos\pi + i\sin\pi$.

$$\therefore\; w = \left\{\cos\left(\frac{2k\pi + \pi}{6}\right) + i\sin\left(\frac{2k\pi + \pi}{6}\right)\right\}, \; k = 0, 1, \ldots, 5$$

$\Rightarrow w = \cos\theta + i\sin\theta$, where $\theta = (2k+1)\dfrac{\pi}{6}$.

From (1), $(2 - z)\,w = 2 + z \Rightarrow z = \dfrac{2(w-1)}{w+1}$

$$\Rightarrow z = \frac{2(\cos\theta + i\sin\theta - 1)}{(\cos\theta + i\sin\theta + 1)}$$

$$= \frac{2\left\{(1 - 2\sin^2\theta/2) + 2i\sin\theta/2\cos\theta/2 - 1\right\}}{\left\{(2\cos^2\theta/2 - 1) + 2i\sin\theta/2\cos\theta/2 + 1\right\}}$$

$$= \frac{2i \sin \theta/2 \{\cos \theta/2 + i \sin \theta/2\}}{\cos \theta/2 \{\cos \theta/2 + i \sin \theta/2\}} = 2i \tan \frac{\theta}{2}$$

Hence $z = 2i \tan \left\{(2k+1)\frac{\pi}{12}\right\}$, $k = 0, 1, 2, ...5.$

Example 23:

Solve $(1 + iz)^{10} + (1 - iz)^{10} = 0.$

Solution :

The given equation can be expressed as

$w^{10} + 1 = 0 \Rightarrow w = (-1)^{1/10} \Rightarrow w = (\cos \pi + i \sin \pi)^{1/10}$,

where $\quad w = \dfrac{1+iz}{1-iz} \Rightarrow z = \dfrac{w-1}{i(w+1)}$

$$= \frac{i(w-1)}{i^2(w+1)} = \frac{i(1-w)}{1+w}.$$

We have

$w = \cos (2k+1)\dfrac{\pi}{10} + i \sin (2k+1)\dfrac{\pi}{10}$, $k = 0, 1, ..., 9.$

$\Rightarrow w = \cos \theta + i \sin \theta$, $\theta = (2k+1)\pi/10$

$$\therefore z = \frac{i(1-w)}{1+w} = \frac{i(1-\cos\theta - i \sin \theta)}{1+\cos\theta + i \sin\theta}$$

$$= \frac{i\{1-(1-2\sin^2\theta/2) - 2i \sin\theta/2 \cos\theta/2\}}{1+2\cos^2\theta/2 - 1 + 2i \sin\theta/2 \cos\theta/2}$$

$$= \frac{-2i^2 \sin\theta/2 (\cos\theta/2 + i \sin\theta/2)}{2\cos\theta/2(\cos\theta/2 + i\sin\theta/2)} = \tan\frac{\theta}{2}$$

$= \tan (2k+1)\dfrac{\pi}{10}$; $k = 0, 1, ...9.$

Hence $z = \pm \tan (p\pi/20)$, $k = 1, 3, 5, 7, 9.$

[$\because$ for $k = 5$, $z = \tan (11\pi/20) = \tan \{\pi - (\pi/20)\} = -\tan (9\pi/20)$ etc.]

Example 24:

***Show that the roots of** $(2 - iz)^8 - z^8 = 0$ **are** $z = -i$, **and** $-i \pm \tan$* $(\pi/8)$*, for* $\pi = 1, 2, 3.$

Solution :

The given equation is expressible as $w^8 - 1 = 0$, where $w = \dfrac{2 - iz}{z}$

$\Rightarrow z = \dfrac{2}{w+i}$.

Now $w^8 = 1 = \cos 0 + i \sin 0 \Rightarrow w = (\cos 2k\pi + i \sin 2k\pi)^{1/8}$

$\therefore w = \cos (k\pi/4) + i \sin (k\pi/4),\ k = 0, 1, ..., 7$

$\Rightarrow w = \cos \theta + i \sin \theta,\ \theta = k\pi/4$

We have $z = \dfrac{2}{w+i} = \dfrac{2}{\cos \theta + i(1+\sin \theta)}$

$$= \frac{2\{\cos \theta - i(1+\sin \theta)\}}{\{\cos \theta + i(1+\sin \theta)\}\{\cos \theta - i(1+\sin \theta)\}}$$

$$= \frac{2\{\cos \theta - i(1+\sin \theta)\}}{\cos^2 \theta + (1+\sin \theta)^2 = 2(1+\sin \theta)}$$

$$= -i + \frac{2 \sin(\pi/4 - \theta/2) \cos(\pi/4 - \theta/2)}{1 + 2\cos^2(\pi/4 - \theta/2) - 1}$$

$$= -i + \tan\left(\frac{\pi}{4} - \frac{\theta}{2}\right) = -i + \tan\left(\frac{\pi}{4} - \frac{k\pi}{8}\right)$$

$-i$ (for $k = 2$); $-i + \tan(2\pi/8)$ (for $k = 0$),

$-i + \tan(\pi/8)$ (for $k = 1$),

$=$ $-i - \tan(\pi/8)$ for ($k = 3$), $-i - \tan(2\pi/8)$ (for $k = 4$),

$-i - \tan(3\pi/8)$ (for $k = 5$), $-i - \infty$ (for $k = 6$),

$-i + \tan(-5\pi/8)$ (for $k = 7$)

$= i - \tan(\pi - 3\pi/8) = -i + \tan(3\pi/8)$.

Hence the roots are $-i$, $-i \pm \tan(p\pi/8)$, $p = 1, 2, 3$, after rejecting the value for $k = 6$.

Example 25:

Prove that

(i) $\cos 5\theta = \cos^5 \theta - 10 \cos^3 \theta \sin^2 \theta + 5 \cos \theta \sin^4 \theta$.

(ii) $\sin 5\theta = 5 \cos^4 \theta \sin \theta - 10 \cos^2 \sin^3 \theta + \sin^5 \theta$.

Solution :

We have $\cos 5\theta + i \sin 5\theta = (\cos \theta + i \sin \theta)^5$

$= \cos^5 \theta + {}^5C_1 \cos^4 \theta \,(i \sin \theta) + {}^5C_2 \cos^3 \theta \,(i \sin \theta)^2$
$\quad + {}^5C_3 \cos^2 \theta \,(i \sin \theta)^3 + {}^5C_4 \cos \theta \,(i \sin \theta)^4 + (i \sin \theta)^5$

$= \cos^5 \theta + 5i \cos^4 \theta \sin \theta - \frac{5.4}{2.1}\, 3 \cos^2 \theta \sin^2 \theta$

$- \frac{5.4.3}{3.2.1}\, i \cos^2 \theta \sin^3 \theta + \frac{5.4.3.2}{4.3.2.1} \cos \theta \sin^4 \theta + i \sin^5 \theta$

$(\because i^2 = -1, i^3 = -i, i^4 = 1, i^5 = i \text{ etc.})$

Thus $\cos 5\theta + i \sin 5\theta = (\cos^5 \theta - 10 \cos^3 \theta \sin^2 \theta + 5 \cos \theta \sin^4 \theta)$
$\quad + i\,(5 \cos^4 \sin \theta - 10 \cos^2 \theta \sin^3 \theta + \sin^5 \theta).$

Equating real and imaginary parts on both the sides, we get

$\cos 5\theta = \cos^5 \theta - 10 \cos^3 \theta \sin^2 \theta + 5 \cos \theta \sin^4 \theta,$

$\sin 5\theta = 5 \cos^4 \theta \sin \theta - 10 \cos^2 \theta \sin^3 \theta + \sin^5 \theta.$

Example 26:

Write down the expansions of cos 6θ, sin 6θ, tan 6θ.

Solution :

$\cos 6\theta + i \sin 6\theta = (\cos \theta + i \sin \theta)^6$

$= \cos^6 \theta + {}^6C_1 \cos^5 \theta \; i \sin\theta + {}^6C_2 \cos^4 \theta \; i^2 \sin^2 \theta + {}^6C_3 \cos^3 \theta \; i^2 \sin^3 \theta$
$\quad + {}^8C_4 \cos^2 \theta \; i^4 \sin^4 \theta + {}^6C_5 \cos \theta \; i^5 \sin^5 \theta + {}^6C_6 \; i^6 \sin^6 \theta.$

Equating real and imaginary parts, we get

$\cos 6\theta = \cos^6 \theta - {}^6C_2 \cos^4 \theta \sin^2 \theta + {}^6C_4 \cos^2 \theta \sin^4 \theta - {}^6C_6 \sin^6 \theta$

$= \cos^6 \theta - 15 \cos^4 \theta \sin^2 \theta + 14 \cos^2 \theta \sin^4 \theta - \sin^6 \theta,$

and $\sin 6\theta = {}^6C_1 \cos^5 \theta \sin \theta - {}^6C_3 \cos^3 \theta \sin^3 \theta + {}^6C_5 \cos \theta \sin^5 \theta$

$= 6 \cos^5 \theta \sin \theta - 20 \cos^3 \theta \sin^3 \theta + 6 \cos \theta \sin^5 \theta.$

Now $\tan 6\theta = \frac{\sin 6\theta}{\cos 6\theta}$

$$= \frac{6 \cos^5 \theta \sin \theta - 20 \cos^3 \theta \sin^3 \theta + 6 \cos \theta \sin^5 \theta}{\cos^6 \theta - 15 \cos^4 \theta \sin^2 \theta + 15 \cos^2 \theta \sin^4 \theta - \sin^6 \theta}$$

On dividing numerator and denominator by $\cos^6 \theta$, we obtain

$$\tan 6\theta = \frac{6 \tan \theta - 20 \tan^3 \theta + 6 \tan^5 \theta}{1 - 15 \tan^2 \theta + 15 \tan^4 \theta - \tan^6 \theta}.$$

Example 27:

Expand cos 7θ and sin 7θ in terms of powers of sin θ and cos θ.

Solution :

By De Moivre's Theorem, we have

$$(\cos\theta + i\sin\theta)^7 = \cos 7\theta + i\sin 7\theta$$

$$\Rightarrow \cos 7\theta + i\sin 7\theta = \cos^7\theta + {}^7C_1\cos^6\theta\,(i\sin\theta) + {}^7C_2\cos^5\theta\,(i\sin\theta)^2 + {}^7C_3\cos^4\theta\,(i\sin\theta)^3 + {}^7C_4\cos^3\theta\,(i\sin\theta)^4 + {}^7C_5\cos^2\theta\,(i\sin\theta)^5 + {}^7C_6\cos\theta\,(i\sin\theta)^6 + {}^7C_7\,(i\sin\theta)^7$$

$$= \cos^7\theta + 7i\cos^6\theta\sin\theta - 21\cos^5\theta\sin^2\theta - 35i\cos^4\theta\sin^3\theta + 35\cos^3\theta\sin^4\theta + 21i\cos^2\theta\sin^5\theta - 7\cos\theta\sin^6\theta - i\sin^7\theta$$

$$= (\cos^7\theta - 21\cos^5\theta\sin^2\theta + 35\cos^3\theta\sin^4\theta - 7\cos\theta\sin^6\theta) + i\,(7\cos^6\theta\sin\theta - 35\cos^4\theta\sin^3\theta + 21\cos^2\theta\sin^5\theta - \sin^7\theta).$$

Equating the real and imaginary parts on both sides, we get

$$\cos 7\theta = \cos^7\theta - 21\cos^5\theta\sin^2\theta + 35\cos^3\theta\sin^4\theta - 7\cos\theta\sin^6\theta,$$

$$\sin 7\theta = 7\cos^6\theta\sin\theta - 35\cos^4\theta\sin^3\theta + 21\cos^2\theta\sin^5\theta - \sin^7\theta.$$

5.3 EXPANSIONS OF COS^N θ AND SIN^N θ IN TERMS OF SINES AND COSINES OF MULTIPLES OF θ, N BEING A POSITIVE INTEGER

Let $z = \cos\theta + i\sin\theta$ so that $\frac{1}{z} = \cos\theta - i\sin\theta$.

Now $z + \frac{1}{z} = 2\cos\theta$ and $z - \frac{1}{z} = 2i\sin\theta$.

By De Moivre's Theorem, we have

$z^n = \cos n\theta + i\sin n\theta$ and $\frac{1}{z^n} = \cos n\theta - i\sin n\theta$.

Thus $z^n + \frac{1}{z^n} = 2\cos n\theta$ and $z^n - \frac{1}{z^n} = 2i\sin n\theta$. ...(2)

From (2), we have

$$(2\cos\theta)^n = \left(z + \frac{1}{z}\right)^n.$$

Expanding the R.H.S. with the help of the Binomial Theorem and using (2), we get the required expansion of $\cos^n\theta$. Similarly

$$(2i\sin\theta)^n = \left(z - \frac{1}{z}\right)^n$$

gives the required expansion of $\sin^n\theta$.

Example 1:

Expand $\sin^7\theta\cos^2\theta$ in a series of sines of multiplies of θ.)

Solution :

Let $z = \cos\theta + i\sin\theta$.

$$\therefore (2i\sin\theta)^7(2\cos\theta)^3 = \left(z-\frac{1}{z}\right)^7\left(z+\frac{1}{z}\right)^3$$

$$= \left(z-\frac{1}{z}\right)^4\left(z^2-\frac{1}{z^2}\right)^3$$

$$= (z^4 - 4z^2 + 6 - 4z^{-2} + z^{-4})(z^6 - 3z^2 + 3z^{-2} - z^{-6})$$

$$= z^{10} - 4z^8 + 3z^6 + 8z^4 - 14z^2 + 14z^{-8} - 8z^{-4} - 3z^{-6} + 4z^{-8} - z^{-10}$$

$$= (z^{10} - z^{-10}) - 4(z^8 - z^{-8}) + 3(z^6 - z^{-6}) + 8(z^4 - z^{-4}) - 14(z^2 - z^{-2})$$

$$= 2i[\sin 10\theta - 4\sin 8\theta + 3\sin 6\theta + 8\sin 4\theta - 14\sin 2\theta].$$

Hence $-2^9\sin^7\theta\cos^2\theta = \sin 10\theta - 4\sin 8\theta + 3\sin 6\theta + 8\sin 4\theta - 14\sin 2\theta.$

Example 2:

Prove that

$-128\sin^6\theta\cos^2\theta = \cos 8\theta - 4\cos 6\theta + 4\cos 4\theta + 4\cos 2\theta - 5.$

Solution :

Let $z = \cos\theta + i\sin\theta$ so that $1/z = \cos\theta - i\sin\theta$.

Thus $z + 1/z = 2\cos\theta$ and $z - 1/z = 2i\sin\theta$.

$$\therefore (2i\sin\theta)^6(2\cos\theta)^2 = \left(z-\frac{1}{z}\right)^6\left(z-\frac{1}{z}\right)^2$$

$$= \left(z-\frac{1}{z}\right)^4\left(z^2-\frac{1}{z^2}\right)^2$$

$$= \left(z^4 - 4z^8.\frac{1}{z} + 6z^2.\frac{1}{z^2} - 4z.\frac{1}{z^3} + \frac{1}{z^4}\right)\left(z^4 - 2 + \frac{1}{z^4}\right)$$

$$= \left(z^8 + \frac{1}{z^8}\right) - 4\left(z^6 + \frac{1}{z^6}\right) + 4\left(z^4 + \frac{1}{z^4}\right) + 4\left(z^2 + \frac{1}{z^2}\right) + 10$$

$$= 2\cos 8\theta - 4(2\cos 6\theta) + 4(2\cos 4\theta) + 4(2\cos 2\theta) - 10$$

$$\Rightarrow -2^8\sin^6\cos^2\theta = 2(\cos 8\theta - 4\cos 6\theta + 4\cos 4\theta + 4\cos 2\theta - 5)$$

Hence $-128 \sin^6\theta\cos^2\theta = \cos 8\theta - 4\cos 6\theta + 4\cos 4\theta + 4\cos 2\theta - 5.$

Example 3:

Prove that

$128 \sin^2\theta \cos^6\theta = -1 \cos 8\theta - 4\cos 6\theta - 4\cos 4\theta + 4\cos 2\theta + 5.$

Solution :

Let $z = \cos\theta + i\sin\theta$ so that $z^{-1} = \cos\theta - i\sin\theta$.

$\therefore z + z^{-1} = 2\cos\theta,\ z - z^{-1} = 2i\sin\theta.$

We have

$(2i\sin\theta)^2 (2\cos\theta)^6 = (z - z^{-1})^2 (z + z^{-1})^6$

$= (z - z^{-1})^2 (z + z^{-1})^2 (z + z^{-1})^4$

$= (z^2 - z^{-2})^2 (z^4 + 4c_1 z^3 z^{-1} + 4c_3 z^2 z^{-2} + 4c_3 z z^{-3} + z^{-4})$

$= (z^4 - 2 + z^{-4})(z^4 + 4z^2 + 6 + 4z^{-2} + z^{-4})$

$= (z^8 + z^{-8}) + 4(z^6 + z^{-6}) + 4(z^4 + z^{-4}) - 4(z^2 + z^{-2}) - 10$

$\therefore -2^8 \sin^2\theta\cos^6\theta = 2\cos 8\theta + 4(2\cos 6\theta) + 4(2\cos 4\theta) - 4(2\cos 2\theta) - 10.$

Hence

$128\sin^2\theta\cos^6\theta = -\cos 8\theta - 4\cos 6\theta - 4\cos 4\theta + 4\cos 2\theta + 5.$

Example 4:

Prove that

$256 \sin^5\theta\cos^4\theta = \sin 9\theta - \sin 7\theta - 4\sin 5\theta + 4\sin 3\theta + 6\sin\theta.$

Solution :

We have

$(2i\sin\theta)^5 (2\cos\theta)^4 = (z - z^{-1})^5 (z + z^{-1})^4$

$= (z^2 - z^{-2})(z - z^{-1})$

$= (z^8 - 4z^4 + 6 - 4z^{-4} + z^{-8})(z - z^{-1})$

$= (z^9 - z^{-9}) - (z^7 - z^{-7}) - 4(z^5 - z^{-5}) + 4(z^3 - z^{-3}) + 6(z - z^{-1})$

$\therefore i\,2^9 \sin^5\theta\cos^4\theta = 2i\sin 9\theta - 2i\sin 7\theta - 4(2i\sin 5\theta) + 4(2i\sin 3\theta) + (2i\sin\theta).$

Hence

$256\sin^5\theta\cos^4\theta = \sin 9\theta - \sin 7\theta - 4\sin 5\theta + 4\sin 3\theta + 6\sin\theta.$

Example 5:

Prove that

(i) $16 \sin^5 \theta = \sin 5\theta - 5 \sin 3\theta + 10 \sin \theta$.

(ii) $16 \cos^5 \theta = \cos 5\theta + 5 \cos 3\theta + 10 \cos \theta$.

Solution :

Let $z = \cos \theta + i \sin \theta$ so that $1/z = \cos \theta - i \sin \theta$.

$$\Rightarrow z - \frac{1}{z} = 2i \sin \theta$$

$$\Rightarrow (2i \sin \theta)^5 = \left(z - \frac{1}{z}\right)^5$$

$$= z^5 - {}^5C_1 z^4 \cdot \frac{1}{z} + {}^5C_2 z^3 \cdot \frac{1}{z^2} - {}^5C_3 z^2 \cdot \frac{1}{z^3} + {}^5C_4 z \cdot \frac{1}{z^4} - \frac{1}{z^5}$$

$$= z^5 - 5 z^3 + \frac{5.4}{2.1} z - \frac{5.4.3}{3.2.1} \cdot \frac{1}{z} + \frac{5.4.3.2}{4.3.2.1} \cdot \frac{1}{z^3} - \frac{1}{z^5}$$

$$= \left(z^5 - \frac{1}{z^5}\right) - 5\left(z^3 - \frac{1}{z^3}\right) + 10\left(z - \frac{1}{z}\right).$$

$= (2i \sin 5\theta) - 5 (2i \sin 3\theta) + 10 (2i \sin \theta)$, by (2) above

$\Rightarrow 32i \sin^5 \theta = 2i (\sin 5\theta - 5 \sin 3\theta + 10 \sin \theta)$ $\quad (\because i^5 = i)$

Hence $16 \sin^5 \theta = \sin 5\theta - 5 \sin 3\theta + 10 \sin \theta$.

(ii) We have $z + 1/z = 2 \cos \theta$, and so

$$(2 \cos \theta)^5 = \left(z + \frac{1}{z}\right)^5 = z^5 + 5z^3 + 10z + 10 \frac{1}{z} + 5 \frac{1}{z^3} + \frac{1}{z^5},$$

(as shown in the first part)

$$= \left(z^5 + \frac{1}{z^5}\right) + 5\left(z^3 + \frac{1}{z^3}\right) + 10\left(z + \frac{1}{z}\right)$$

$\Rightarrow 2^5 \cos^5 \theta = (2 \cos 5\theta) + 5 (2 \cos 3\theta) + 10 (2 \cos \theta)$, by (2) above.

Hence $16 \cos^5 \theta = \cos 5\theta + 5 \cos 3\theta + 10 \cos \theta$.

Example 6:

Prove that

$32 \cos^6 \theta = \cos 6\theta + 6 \cos 4\theta + 15 \cos 2\theta + 10$.

Solution :

Let $x = \cos \theta + i \sin \theta$ so that $1/x = \cos \theta - i \sin \theta$.

Thus $2\cos\theta = x + \frac{1}{x} \Rightarrow (2\cos\theta)^6 = \left(x+\frac{1}{x}\right)^6$

$$= x^6 + {}^6C_1x^5 \cdot \frac{1}{x} + {}^6C_2x^4 \cdot \frac{1}{x^2} + {}^6C_3x^3 \cdot \frac{1}{x^3} + {}^6C_4x^2 \cdot \frac{1}{x^4} + {}^6C_5x \cdot \frac{1}{x^5} + \frac{1}{x^6}$$

$$= x^6 + 6x^4 + 15x^2 + +20 + \frac{15}{x^2} + \frac{6}{x} + \frac{1}{x^6}$$

$$= \left(x^6 + \frac{1}{x^6}\right) + 6\left(x^4 + \frac{1}{x^4}\right) + 15\left(x^2 + \frac{1}{x^2}\right) + 20$$

$\therefore 64\cos^6\theta = 2\cos 6\theta + 6.2\cos 4\theta + 15.2\cos 2\theta + 20$

Hence $32\cos^6\theta = \cos 6\theta + 6\cos 4\theta + 15\cos 2\theta + 10.$

Example 7:

Prove that

$$64\cos^7\theta = \cos 7\theta + 7\cos 5\theta + 21\cos 3\theta + 35\cos\theta.$$

Solution :

Let $x = \cos\theta + i\sin\theta$, then $1/x = \cos\theta - i\sin\theta$.

$$\therefore (2\cos\theta)^7 = \left(x+\frac{1}{x}\right)^7 = x^7 + 7c_1.x^6 \cdot \frac{1}{x} + 7\,c_2.x^5 \cdot \frac{1}{x^2}$$

$$+ 7c_3x^4.\frac{1}{x^3} + 7c_4.x^3.\frac{1}{x^4} + 7c_5x^2. + 7c_6x.\frac{1}{x^6} + 7c_7.\frac{1}{x^7}$$

$$= x^7 + 7x^5 + 21x^3 + 35x + 35.\frac{1}{x} + 21.\frac{1}{x^3} + 7.\frac{1}{x^5} + \frac{1}{x^7}$$

$$= \left(x^7 + \frac{1}{x^7}\right) + 7\left(x^5 + \frac{1}{x^5}\right) + 21\left(x^3 + \frac{1}{x^3}\right) + 35\left(x + \frac{1}{x}\right)$$

$= 2\,[\cos 7\theta + 7\cos 5\theta + 21\cos 3\theta + 35\cos\theta]$.

Hence $64\cos^7\theta = \cos 7\theta + 7\cos 5\theta + 21\cos 3\theta + 35\cos\theta$.

Example 8:

Prove that

$$\sin^8\theta = \frac{1}{128}\{\cos 8\theta - 8\cos 6\theta + 28\cos 4\theta - 56\cos 2\theta + 70\}.$$

Solution :

Let $z = \cos\theta + i\sin\theta$

$\Rightarrow z^{-1} = \cos\theta - i\sin\theta$

$\Rightarrow z - z^{-1} = 2i\sin\theta$

$$\Rightarrow \sin^8\theta = \frac{1}{(2i)^8}(z - z^{-1})^8$$

$$= \frac{1}{256}(z^8 - {}^8c_1 z^6 + {}^8c_2 z^4 - {}^8c_3 z^2 + {}^8c_4 - {}^8c_5 z^{-2} + {}^8c_6 z^{-4} - {}^8c_7 z^{-6} + z^{-8})$$

$$= \frac{1}{256}\{(z^8 + z^{-8}) - 8(z^6 + z^{-6}) + 28(z^4 + z^{-4}) - 56(z^2 + z^{-2}) + 70\}$$

$$= \frac{1}{128}\{\cos 8\theta - 8\cos 6\theta + 28\cos 4\theta - 56\cos 2\theta + 70\}.$$

Example 9:

Show that

$$\sin^6\theta = -\frac{1}{32}[\cos 6\theta - 6\cos 4\theta + 15\cos 2\theta - 10].$$

Solution :

We have

$$(2i\sin\theta)^6 = \left(x - \frac{1}{x}\right)^6$$

$$= x^6 + 6c_1 x^5\left(-\frac{1}{x}\right) + 6c_2 x^4\left(-\frac{1}{x}\right)^2$$

$$+ 6c_3 x^3\left(-\frac{1}{x}\right)^3 + 6c_4 x^2\left(-\frac{1}{x}\right)^4 + 6c_5 x^2\left(-\frac{1}{x}\right)^5 + 6c_6\left(-\frac{1}{x}\right)^6$$

$$= x^6 - 6x^4 + 15x^2 - 20 + 15\frac{1}{x^2} - 6\frac{1}{x^4} + \frac{1}{x^6}$$

$$= \left(x^6 + \frac{1}{x^6}\right) - 6\left(x^4 + \frac{1}{x^4}\right) + 15\left(x^2 + \frac{1}{x^2}\right) - 20$$

$$= [2\cos 6\theta] - 6[2\cos 4\theta] + 15[2\cos 2\theta] - 20$$

Hence $\sin^6\theta = -\frac{1}{32}[\cos 6\theta - 6\cos 4\theta + 15\cos 2\theta - 10]$

Example 10:

Prove that

$$32 \sin^4 \theta \cos^2 \theta = \cos 6\theta - 2 \cos 4\theta - \cos 2\theta + 2.$$

Solution :

Let $x = \cos \theta + i \sin \theta$

$\Rightarrow 1/x = \cos \theta - i \sin \theta.$

$\Rightarrow x + 1/x = 2 \cos \theta$ and $x - 1/x = 2i \sin \theta,$

$$\Rightarrow (2i \sin \theta)^4 (2 \cos \theta)^2 = \left(x+\frac{1}{x}\right)^2 \left(x-\frac{1}{x}\right)^4$$

$$= \left(x+\frac{1}{x}\right)^2 \left(x-\frac{1}{x}\right)^2 \left(x-\frac{1}{x}\right)^2$$

$$= \left(x^2-\frac{1}{x^2}\right)^2 \left(x-\frac{1}{x}\right)^2$$

$$= \left(x^4-2+\frac{1}{x^4}\right)\left(x^2-2+\frac{1}{x^2}\right)$$

$$= x^6 - 2x^4 - x^2 + 4 - \frac{1}{x^2} - \frac{2}{x^4} + \frac{1}{x^6}$$

$$= \left(x^6+\frac{1}{x^6}\right) - 2\left(x^4+\frac{1}{x^4}\right) - \left(x^2+\frac{1}{x^2}\right) + 4$$

$$= 2 \cos 6\theta - 2.2 \cos 4\theta - 2 \cos 2\theta + 4.$$

Hence $32 \sin^4 \theta \cos^2 \theta = \cos 6\theta - 2 \cos 4\theta - \cos 2\theta + 2.$

Example 11:

Prove that

$$128 \cos^3 \theta \sin^5 \theta = \sin 8\theta - 2 \sin 6\theta - 2 \sin 4\theta + 6 \sin 2\theta.$$

Solution :

If $z = \cos \theta + i \sin \theta$, then $z^{-1} = \cos \theta - i \sin \theta.$

$\Rightarrow 2 \cos \theta = z + z^{-1}, 2i \sin \theta = z - z^{-1}.$

$$\Rightarrow (2 \cos \theta)^3 (2i \sin \theta)^5 = (z + z^{-1})^3 (z - z^{-1})^5$$

$$= (z^2 - z^{-2})^3 (z - z^{-1})^2$$

$$= (z^6 - 3z^2 + 3z^{-2} - z^{-6}) (z^2 - 2 + z^{-2})$$

$$= (z^8 - z^{-8}) - 2 (z^6 - z^{-6}) - 2 (z^4 - z^{-4}) + 6 (z^2 - z^{-2})$$

$2^8\, i \cos^3\theta \sin^5\theta = 2i \sin 8\theta - 2\,(2i \sin 6\theta) - 2\,(2i \sin 4\theta)$
$+ 6\,(2i \sin 2\theta)$

Hence $128 \cos^3\theta \sin^5\theta = \sin 8\theta - 2 \sin 6\theta - 2 \sin 4\theta + 6 \sin 2\theta.$

Example 73:

Prove that

$64 \sin^3\theta \cos^4\theta = -\sin 7\theta - \sin 5\theta + 3 \sin 3\theta + 3 \sin \theta.$

Solution :

$(2\cos\theta)^4 (2i \sin\theta)^3 = (z + z^{-1})^4 (z - z^{-1})^3$

$= (z^4 + 4z^2 + 6 + 4z^{-2} + z^{-4})(z^3 - 3z + 3z^{-1} - z^{-3})$

$= (z^7 - z^{-7}) + (z^5 - z^{-5}) - 3\,(z^3 - z^{-3}) - 3\,(z - z^{-1})$

$\Rightarrow -2i\,(64 \sin^3\theta \cos^4\theta) = 2i \sin 7\theta + 2i \sin 5\theta - 3\,(2i \sin 3\theta)$
$- 3\,(2i \sin\theta)$

Hence $64 \sin^2\theta \cos^4\theta = -\sin 7\theta - \sin 5\theta + 3 \sin 3\theta + 3 \sin\theta.$